EXTENSIONS IN MATHEMATICS™ SERIES
C

EXTENSIONS IN MATHEMATICS

- ☐ PROVIDES CHALLENGING INSTRUCTIONAL ACTIVITIES FOR 12 MATHEMATICS STRATEGIES
- ☐ STRENGTHENS PROBLEM-SOLVING SKILLS AND IMPROVES MATH-RELATED WRITING SKILLS
- ☐ FEATURES ASSESSMENT IN MATHEMATICS, INCLUDING SELECTED-RESPONSE AND CONSTRUCTED-RESPONSE PROBLEMS

CURRICULUM ASSOCIATES®, INC.

Acknowledgments

Product Development and Design by Chameleon Publishing Services
   Written by Susan A. DeStefano and Pamela Halloran
   Illustrated by Leslie Alfred McGrath

ISBN 978-0-7609-3697-9
©2006, 2004—Curriculum Associates, Inc.
North Billerica, MA 01862

No part of this book may be reproduced by any means
without written permission from the publisher.
All Rights Reserved. Printed in USA.

15 14 13 12 11 10 9 8 7 6 5 4

Table of Contents

STRATEGY ONE	Building Number Sense	4
STRATEGY TWO	Using Estimation	14
STRATEGY THREE	Applying Addition	24
STRATEGY FOUR	Applying Subtraction	34
STRATEGY FIVE	Applying Multiplication	44
STRATEGY SIX	Applying Division	54
STRATEGY SEVEN	Converting Time and Money	64
STRATEGY EIGHT	Converting Customary and Metric Measures	74
STRATEGY NINE	Using Algebra	84
STRATEGY TEN	Using Geometry	94
STRATEGY ELEVEN	Determining Probability and Averages	104
STRATEGY TWELVE	Interpreting Graphs and Charts	114
STRATEGIES ONE–TWELVE REVIEW		124

STRATEGY ONE: Building Number Sense

Learn About Number Sense

Thinking about the strategy

Every numeral in a number has a place value. How would you use digits to write these numbers?
- seven thousand, six hundred fifty-two and six tenths
- four thousand, forty-four and twenty-five hundredths
- two thousand, five and seventy-nine hundredths

Each number has thousands. Each number has a part that is less than 1. To write such a number in standard form, you can use decimals.

Lynn used a place-value chart to write the numbers above in standard form. Where did she write each digit? What mark did she use to separate thousands from hundreds? What mark did she use to show tenths and hundredths?

= 1 or more				= less than 1	
thousands (1,000)	hundreds (100)	tens (10)	ones (1)	tenths (0.1)	hundredths (0.01)
7,	6	5	2	.6	
4,	0	4	4	.2	5
2,	0	0	5	.7	9

Studying the problem Read the problem and the notes beside it.

What was the total in bills and coins?

What was the total in checks?

What was the total in credit-card receipts?

Last Saturday, Dean helped his dad count what was in their store's cash register at closing. They counted one thousand, three hundred eighty-two dollars and sixty-three cents in bills and coins; two thousand, one hundred five dollars and fifty cents in checks; and three thousand, two hundred fifty dollars and eight cents in credit-card receipts. How are these amounts written in standard form?

How can Dean use a place-value chart to solve the problem?

Studying the solution

A **place-value chart** is a graphic organizer that you can use to understand and write large numbers with decimals. Dean used this place-value chart to show how much money in cash, checks, and credit-card receipts were in the cash register at the end of the day.

thousands (1,000)	hundreds (100)	tens (10)	ones (1)	tenths (0.1)	hundredths (0.01)
1,	3	8	2	.6	3
2,	1	0	5	.5	0
3,	2	5	0	.0	8

Dean showed the amounts as $1,382.63; $2,105.50; and $3,250.08.

Understanding the solution

Read what Dean wrote to explain how he used a place-value chart to solve the problem.

Each digit in a number has a place value. A digit's value depends on its place in the number. Digits to the left of the decimal point can have a place value of ones, tens, hundreds, thousands, and so on. Digits to the right of the decimal point have a place value of tenths, hundredths, and so on.

To show one thousand, three hundred eighty-two dollars and sixty-three cents, I wrote 1 in the thousands column, 3 in the hundreds column, 8 in the tens column, 2 in the ones column, 6 in the tenths column, and 3 in the hundredths column. I put a comma after the 1 to separate thousands from hundreds. I put a decimal point in front of the 6 to show tenths and hundredths.

I used the same steps to write $2,105.50 and $3,250.08.

Solve a Problem

Studying the problem Read the problem. As you read, think about how you could use a place-value chart to solve the problem.

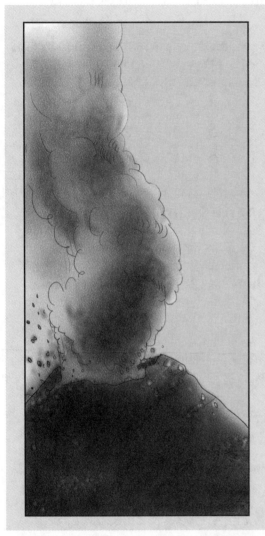

On a recent visit to the library, Aretha used an online encyclopedia to find the height in meters of some of the world's most famous mountains.

Aretha learned that Mount Kilimanjaro, the highest mountain in Africa, is five thousand, eight hundred ninety-one and eight tenths meters high. Mount Everest, the highest mountain in the world, is eight thousand, eight hundred forty-nine and forty-four hundredths meters high. Mount Fuji, on Honshu Island in Japan, is three thousand, seven hundred seventy-five and sixty-eight hundredths meters high. Italy's Mount Vesuvius, which is the only active volcano on Europe's mainland, is one thousand, two hundred seventy-seven and five hundredths meters high.

How is the height of each mountain written in digits?

Finding the solution Use the information from the problem to complete this place-value chart. Then write your answer below.

thousands (1,000)	hundreds (100)	tens (10)	ones (1)	tenths (0.1)	hundredths (0.01)

Answer: _____

Explaining the solution

Reread "Understanding the solution" on page 5, which tells how Dean used a place-value chart to solve a problem. Then write your own explanation of how you completed the place-value chart on page 6 and found your solution.

_____

_____

_____

_____

_____

_____

_____

Applying the solution

Use your place-value chart on page 6 to answer these questions.

1. How many of the numbers in your place-value chart have parts that equal less than 1? _____

2. What digit should be in the tenths column for the height of Mount Vesuvius? _____

3. For the height of Mount Fuji, between what two digits is the decimal point? _____

4. What is the value of the 8 in the height of Mount Fuji?

   _____

5. Mount St. Helens, an active volcano in Washington, is two thousand, five hundred forty-nine and twenty-two hundredths meters high. If this figure was included in the chart, what digits would be in the thousands, tenths, and hundredths places?

   _____

Learn More About Number Sense

Thinking about the strategy

All the numbers below are mixed numbers. A mixed number is a whole number and a fraction together.

$1\frac{1}{2}$

$3\frac{2}{3}$

$2\frac{3}{4}$

Geometric figures such as squares, circles, triangles, and other shapes are graphic organizers that you can use to show mixed numbers.

How can Marty find if $2\frac{3}{4}$ is closer to the whole number 2 or to the whole number 3? Marty used these squares to find the answer.

$2\frac{3}{4}$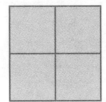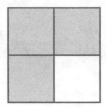

Answer: Marty found that $2\frac{3}{4}$ is closer to 3 than to 2.

Understanding the solution

Read what Marty wrote to explain how he used squares to solve the problem.

Figure out how many whole shapes are needed.

Divide each shape into the same number of parts as the denominator.

Color the whole shapes. Color the correct fraction of the shape.

First, I looked at the mixed number. The number $2\frac{3}{4}$ is between 2 and 3, so I needed 2 whole squares plus part of a third square. Since the denominator in the fraction is 4, I divided each square into 4 parts, with each part equal to $\frac{1}{4}$. To show the whole number 2, I colored all the parts in 2 squares. To show the fraction $\frac{3}{4}$, I colored 3 parts of the third square. Since I could see that more than half of the parts in the third square were colored, I could tell that $2\frac{3}{4}$ is closer to the whole number 3 than to the whole number 2.

Solve a Problem

Finding the solution

Leon wants to make pizza for dinner. The recipe calls for $3\frac{5}{8}$ cups of flour.

Is the amount that Leon needs closer to 3 cups or 4 cups?

Complete the dividing and coloring of these circles to find whether the amount is closer to 3 cups or 4 cups. Then write your answer below.

Answer: _____

Explaining the solution

Write an explanation of how you found the answer by completing the circles above.

_____

_____

_____

_____

_____

_____

_____

_____

_____

Numbers in Context

Read "Library Repairs." Think about the ways that numbers are used in the selection. Then answer items A–C on page 11.

Library Repairs

Kathy lives next door to a public library. One Saturday, Kathy looked out her window and saw people running in and out of the library, carrying piles and piles of books. They were loading the books into the trunks of their cars.

Kathy ran into the kitchen. "Mom," she said, "people are stealing all the books from the library."

"No, they're not," Kathy's mom said. "The pipes burst during the night and flooded the children's library. People are helping to rescue the books. I'm going to help, too. Do you want to come?"

Kathy, her mom, and dozens of neighbors spent the rest of the day carting books and furniture out of the children's library. At the end of the day, all that was left in the basement room was a soggy carpet and empty metal shelves. It would take days, maybe even weeks, for the room to dry out.

The newspaper printed an article about the flood, along with what the repair costs would be. The plumber's bill would be two thousand, seven hundred fifty dollars. The carpet, which had been ruined, had been pulled up. It was to be replaced at a cost of one thousand, two hundred thirty-seven dollars and ten cents. The metal bookshelves were not damaged. However, more than 350 books were destroyed. Mr. Olivetti, the librarian, figured out that it would cost four thousand, twelve dollars and sixty-two cents to replace those books.

While the repair work was being done, the children's library was put into huge storage containers. After two months, repairs were completed, and the books were moved back into the library. Kathy watched from her window as each container of books was unloaded. Every inch of the first 4 containers was packed with books. Only $\frac{2}{3}$ of the last container was filled.

A. How should each repair cost be written in standard form? Use the information from page 10 to complete this place-value chart. Then write your answer below.

thousands (1,000)	hundreds (100)	tens (10)	ones (1)	tenths (0.1)	hundredths (0.01)

Answer: _____

B. Did the books fill closer to 4 or to 5 storage containers? Use the information from page 10 to divide and color these rectangles. Then write your answer below.

Answer: _____

C. Explain your solution to either item A or item B above.

_____

_____

_____

_____

_____

_____

_____

Check Your Understanding

Fill in the letter of the correct answers to questions 1–8.
Write your answers to questions 9 and 10.

1. Pat wrote the number three thousand, two hundred forty-nine and six tenths in standard form. Which of these should Pat have written?
   - Ⓐ 324.96
   - Ⓑ 3,249.6
   - Ⓒ 3,249.06
   - Ⓓ 32,496

2. Elaine's mom wrote a check for $5,090.80. Which of these shows the amount in words?
   - Ⓐ five thousand, nine hundred eighty dollars
   - Ⓑ five thousand, nine dollars and eight cents
   - Ⓒ five thousand, ninety dollars and eighty cents
   - Ⓓ five thousand, nine dollars and eighty cents

3. On Monday, highway workers in Mike's town repaired a stretch of road 1,974.08 feet long. What is the value of the 8 in 1,974.08 feet?
   - Ⓐ eight hundredths of a foot
   - Ⓑ eight hundred feet
   - Ⓒ eight feet
   - Ⓓ eight tenths of a foot

4. Juan's soccer team collected $1,365.72 at their yearly bake sale and car wash. Which of the following has the greatest value in the number 1,365.72?
   - Ⓐ 6
   - Ⓑ 2
   - Ⓒ 7
   - Ⓓ 1

5. Kip and his dad are building a model sailboat. The plans call for a piece of wood $3\frac{3}{8}$ inches long. What amount is closest to what they need?
   - Ⓐ 2 inches
   - Ⓑ 3 inches
   - Ⓒ 4 inches
   - Ⓓ 5 inches

6. Chan's mom needs $3\frac{1}{4}$ gallons of juice to make a party punch recipe. The juice comes in gallon cans. Does Chan's mom need closer to 1, 2, 3, or 4 cans of juice?
   - Ⓐ 1 can
   - Ⓑ 2 cans
   - Ⓒ 3 cans
   - Ⓓ 4 cans

7. Tasha wrote a mixed number that was closer to 6 than to 5 or 7. Which of these numbers did she write?
   - Ⓐ $5\frac{1}{3}$
   - Ⓑ $6\frac{1}{4}$
   - Ⓒ $6\frac{3}{4}$
   - Ⓓ $7\frac{2}{5}$

8. Luke collects action figures. He displays them in a special case with shelves that hold 10 figures each. So far, he has 4 complete rows of figures and 1 row with 3 figures. What mixed number shows how much of the shelves have been filled?
   - Ⓐ $4\frac{3}{10}$
   - Ⓑ $3\frac{1}{2}$
   - Ⓒ $4\frac{1}{3}$
   - Ⓓ $5\frac{3}{10}$

9. Wilson helped his grandmother count her large collection of buttons, which she keeps in tins. One tin had 1,000 white buttons. Another tin had 400 black and brown buttons. Another tin had 70 cloth-covered buttons. A fourth tin had 8 buttons shaped like wild animals. How many buttons did Wilson's grandmother have all together? Write your answer in standard form.

10. Phil made sandwiches for lunch. Phil and Jodie each ate 1 sandwich. Matt ate $1\frac{1}{4}$ sandwiches. Is the number of sandwiches that the friends ate closer to 3 sandwiches or 4 sandwiches? Explain how you found your answer.

Extend Your Learning

- *Mix and Stir*

    Demo's Diner serves pie slices for dessert. For today, the diner has more than 3 pies but fewer than 7 pies. Draw and label two pictures that show how many pies the diner could have. Each picture must show a mixed number.

- *Social Studies: In High Places*

    With a partner, find the height in meters of the following mountains: Mount Olympus in Greece, Mount McKinley in Alaska, and Mount Kenya in Kenya, Africa. Write the heights in standard form.

STRATEGY TWO: Using Estimation

Learn About Estimation

Thinking about the strategy

You can use estimation to find the ten or hundred closest to a number. You could use estimation to find about how many pages are in a book. Or about how many days until your birthday.

To estimate, you round to find a number that is close to an exact number. You round up if a digit is 5 or more. You round down if a digit is 4 or less.

Using a number line can help you round a number. How did Zoey round 36 to the nearest ten and 438 to the nearest hundred?

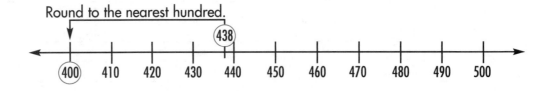

Studying the problem Read the problem and the notes beside it.

How many people will be at the dinner?

How many awards will be given out?

What place should you look at to find the nearest ten? The nearest hundred?

Ricardo's dad coaches the high-school swim team. At the end of every season, Ricardo helps his dad plan an awards dinner for swimmers and their families. This year, 210 people will be at the dinner, and at least 25 awards will be given out.

To the nearest ten, about how many award certificates need to be printed? To the nearest hundred, about how many places need to be set for the dinner?

How can Ricardo use a number line to solve the problem?

Studying the solution   A **number line** is a graphic organizer that you can use to round numbers to the nearest ten and the nearest hundred. Ricardo made these number lines. He used them to figure out about how many award certificates have to be printed and about how many places have to be set for the dinner.

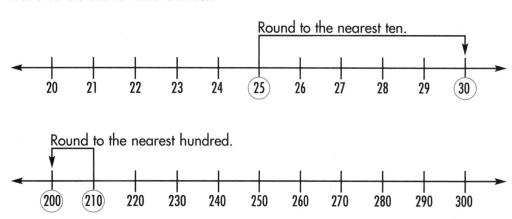

Ricardo figured out that about 30 award certificates need to be printed and about 200 places need to be set.

Understanding the solution   Read what Ricardo wrote to explain how he used number lines to solve the problem.

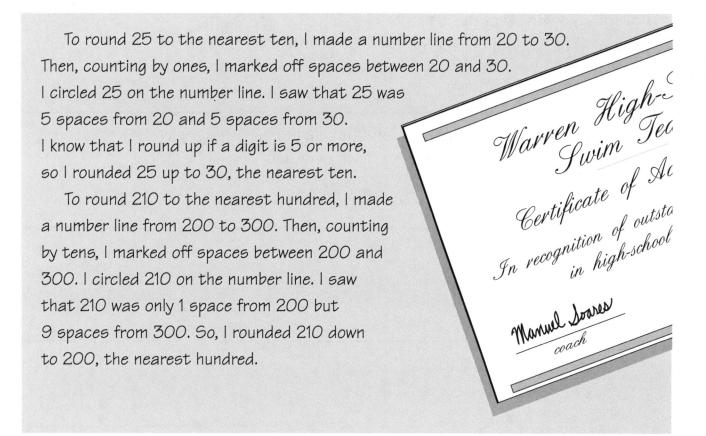

To round 25 to the nearest ten, I made a number line from 20 to 30. Then, counting by ones, I marked off spaces between 20 and 30. I circled 25 on the number line. I saw that 25 was 5 spaces from 20 and 5 spaces from 30. I know that I round up if a digit is 5 or more, so I rounded 25 up to 30, the nearest ten.

To round 210 to the nearest hundred, I made a number line from 200 to 300. Then, counting by tens, I marked off spaces between 200 and 300. I circled 210 on the number line. I saw that 210 was only 1 space from 200 but 9 spaces from 300. So, I rounded 210 down to 200, the nearest hundred.

15

Solve a Problem

Studying the problem Read the problem. As you read, think about how you could use a number line to solve the problem.

Ming's grandfather has always enjoyed bird-watching. Many years ago, he began keeping track of new bird sightings. Whenever he sees a bird he has never seen before, he writes about it in a journal. He writes the date, the time, the place, and the name of the bird. If he doesn't know the name of a bird, he describes it or he draws a picture. Then he goes home and looks for the bird's name in one of his many bird books.

Ming loves to flip through her grandfather's 18 journals. Her grandfather has notes on 750 kinds of birds. Ming especially likes to look at the drawings. Some are quick sketches. Some are so detailed and colorful that the birds look as if they could fly off the page.

To the nearest ten, how many journals has Ming's grandfather filled? To the nearest hundred, how many kinds of birds has he written about in those journals?

Finding the solution Use the information from the problem to complete these number lines. Then write your answer below.

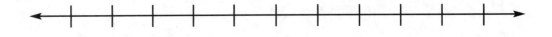

Answer: _____

16

Explaining the solution

Reread "Understanding the solution" on page 15, which tells how Ricardo used number lines to solve a problem. Then write your own explanation of how you completed the number lines on page 16 and found your solution.

_____

_____

_____

_____

_____

_____

_____

Applying the solution

Use your number lines on page 16 to answer these questions.

1. What two numbers did you use at either end of the number line to estimate the number of journals? _____

2. Did you round up or down to estimate the number of journals?

   _____

3. What two numbers did you use at either end of the number line to estimate how many birds were described in the journals?

   _____

4. Did you round up or down to estimate the number of birds described in the journals? _____

5. In one journal, Ming's grandfather described 48 birds. To the nearest ten, about how many birds did he describe in that journal?

   _____

Learn More About Estimation

Thinking about the strategy

You may sometimes want to estimate a sum of money. First, round each money amount to the nearest dollar or to the nearest ten cents. Then add the rounded amounts together to find the estimated sum.

A **table** is a graphic organizer that you can use to estimate a sum.

How can Arthur find the estimated sums of $2.64 and $1.16 and of $0.12 and $0.55? Arthur used this table to find the answer.

To round to the nearest dollar, look at the number of dimes. If it is 5 or more, round up. If less than 5, round down.

Dollars (1 dollar = 100 cents)	Dimes (1 dime = 10 cents)	Pennies (1 penny = 1 cent)	Rounded amounts to the nearest dollar
2	⑥	4	$3.00
1	①	6	+ $1.00
		Estimated Sum	**$4.00**

To round to the nearest ten cents, look at the number of pennies. If it is 5 or more, round up. If less than 5, round down.

Dollars (1 dollar = 100 cents)	Dimes (1 dime = 10 cents)	Pennies (1 penny = 1 cent)	Rounded amounts to the nearest ten cents
0	1	②	$0.10
0	5	⑤	+ $0.60
		Estimated Sum	**$0.70**

Answer: The estimated sum of $2.64 and $1.16 is $4.00; the estimated sum of $0.12 and $0.55 is $0.70.

Understanding the solution

Read what Arthur wrote to explain how he used a table to solve the problem.

The digit before the decimal tells how many dollars.

The first digit after the decimal tells how many dimes.

The second digit after the decimal tells how many pennies.

To find the estimated sum of $2.64 and $1.16, I broke down each amount into dollars, dimes, and pennies. Then I looked at the number of dimes in each amount. There were 6 dimes in $2.64, so I rounded up to $3.00. There was 1 dime in $1.16, so I rounded down to $1.00. I added $3.00 and $1.00 to get the estimated sum of $4.00. To find the estimated sum of $0.12 and $0.55, I rounded $0.12 down to $0.10 and $0.55 up to $0.60. I added and I found the estimated sum to be $0.70.

Solve a Problem

Finding the solution

At the grocery store today, Jenny's mom spent $2.19 for a loaf of bread, $2.65 for a gallon of milk, and $5.40 for a roasting chicken. Jenny learned that those same items would have cost $0.11, $0.13, and $0.27 in 1903. To the nearest dollar, about how much money did Jenny's mom spend at the grocery store? To the nearest ten cents, about how much would those same items have cost in 1903?

Complete this table to find the estimated sums. Then write your answer below.

Dollars (1 dollar = 100 cents)	Dimes (1 dime = 10 cents)	Pennies (1 penny = 1 cent)	Rounded amounts to the nearest dollar
		Estimated Sum	

Dollars (1 dollar = 100 cents)	Dimes (1 dime = 10 cents)	Pennies (1 penny = 1 cent)	Rounded amounts to the nearest ten cents
		Estimated Sum	

Answer: _____

Explaining the solution

Write an explanation of how you found the answer by completing the table above.

_____
_____
_____
_____
_____

Numbers in Context

Read "The Bookstore." Think about the ways that numbers are used in the selection. Then answer items A–C on page 21.

The Bookstore

Irma and her mom walked into the bookstore. The small, cramped store was almost dark. Piles of used books—on shelves, on chairs, and on the floor—surrounded them. "Whatever we're looking for," Irma whispered to her mom, "we'll never find it in this mess."

"Irma!" her mom said. "The store owner might hear you."

At that moment, an elderly man shuffled through a pair of heavy black curtains at the back of the shop.

"Can I help you?" he asked with a warm smile.

"I'm looking for a special book of fairy tales," Irma's mom said. "This book has a red cloth binding. The pages have gold on the edges. And there's a picture of the Ugly Duckling on the cover. The book contains 45 fairy tales by Hans Christian Andersen."

"He published 168 fairy tales in all, you know," the man informed Irma and her mom. "May I ask why this book is so special to you?"

"My grandmother gave me a copy of the book for my sixth birthday," Irma's mom said. "I treasured it. I must have lost it when I left home for college. I've been searching for another copy for years."

"Wait here," the man said. Moments later, he returned with several books in his hands. One was exactly like the book Irma's mom had described. There was a price sticker on it that said $6.88.

Irma's mom gasped with delight. "It's just like the book my grandmother gave me," she said. Then she took the book from the man and opened the front cover. On the front page were the words, "To my dear granddaughter for her sixth birthday. Happy reading."

"I'll take it," she said.

A. To the nearest ten, about how many fairy tales were in the book that Irma's mom had lost? To the nearest hundred, about how many fairy tales did Hans Christian Andersen publish in all? Use the information from page 20 to complete these number lines. Then write your answer below.

Answer: _____

B. Irma's mom also bought a used copy of *Grimm's Fairy Tales*. It cost $8.25. To the nearest dollar, how much did Irma's mom pay for both books? Use the information from page 20 to complete this table. Then write your answer below.

Dollars (1 dollar = 100 cents)	Dimes (1 dime = 10 cents)	Pennies (1 penny = 1 cent)	Rounded amounts to the nearest dollar
		Estimated Sum	

Answer: _____

C. Explain your solution to either item A or item B above.

_____

_____

_____

_____

_____

_____

Check Your Understanding

Fill in the letter of the correct answers to questions 1–8.
Write your answers to questions 9 and 10.

1. Miko broke her arm and had to wear a cast for 64 days. To the nearest ten, about how many days did Miko have to wear the cast?
   - Ⓐ 65 days
   - Ⓑ 70 days
   - Ⓒ 60 days
   - Ⓓ 75 days

2. Vivian has about 300 baseball cards in her collection. What number, when rounded to the nearest hundred, is 300?
   - Ⓐ 351
   - Ⓑ 251
   - Ⓒ 367
   - Ⓓ 219

3. A series of winter storms dumped about 90 inches of snow in the mountains. What number, when rounded to the nearest ten, is 90?
   - Ⓐ 83
   - Ⓑ 85
   - Ⓒ 97
   - Ⓓ 95

4. There are 574 students in Paco's school. To the nearest hundred, about how many students are in Paco's school?
   - Ⓐ 600 students
   - Ⓑ 550 students
   - Ⓒ 500 students
   - Ⓓ 570 students

5. Eli bought two small toy cars at a yard sale. He paid $0.19 for one and $0.24 for the other. To the nearest ten cents, about how much did Eli pay for both cars?
   - Ⓐ $0.50
   - Ⓑ $0.45
   - Ⓒ $0.40
   - Ⓓ $0.30

6. Katy and her sister bought two used movie DVDs at the video store for $6.15 each. To the nearest dollar, about how much did they spend for the two DVDs?
   - Ⓐ $12.50
   - Ⓑ $13.00
   - Ⓒ $12.20
   - Ⓓ $12.00

7. Wendy and her mom rode the subway home. Wendy's fare was $0.35. Her mom's fare was $0.75. To the nearest ten cents, how much did Wendy and her mom pay together to ride the subway?
   - Ⓐ $1.00
   - Ⓑ $1.40
   - Ⓒ $1.20
   - Ⓓ $1.30

8. Doug bought three pencils for $0.57, a small notebook for $0.49, and a plastic ruler for $0.33. To the nearest ten cents, about how much did Doug spend?
   - Ⓐ $1.40
   - Ⓑ $1.20
   - Ⓒ $1.50
   - Ⓓ $1.00

9. Tish read that there are 740 different kinds of tree frogs in the world. To the nearest hundred, about how many kinds of tree frogs are there?

10. Dylan's dad took Dylan and Zach to lunch. Dylan and Zach each had the children's special. Dylan's dad had a salad and a glass of iced tea. Dylan's and Zach's lunches came to $4.06 each. Dylan's dad's lunch came to $8.91. To the nearest dollar, about how much did Dylan's dad pay for all three lunches? Explain how you found your answer.

Extend Your Learning

- *How Many Days?*

    Find, to the nearest ten or nearest hundred, about how many days it is until: your birthday; school vacation; summer vacation; your favorite holiday. Compare your estimates with those of a partner.

- *Art: It's a Colorful World*

    Look through an art-supply catalog. Make a list of different-size boxes of crayons and their prices. Then find, to the nearest dollar, about how much it would cost to buy the smallest- and the largest-size boxes.

STRATEGY THREE
Applying Addition

Learn About Addition

Thinking about the strategy

You add numbers to find the sum. You can add numbers with four digits as easily as you add numbers with two or three digits. A four-digit number has a digit in the ones place, tens place, hundreds place, and thousands place. To add four-digit numbers, line up the digits according to place value. Then add from right to left. Regroup when a column adds up to 10 or more.

A flowchart can help you add numbers with four digits, by guiding you through each step of addition.

What is the sum of 2,196 and 3,645? After you add each column, you may need to regroup.

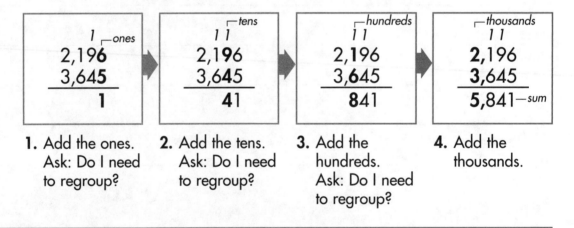

1. Add the ones. Ask: Do I need to regroup?
2. Add the tens. Ask: Do I need to regroup?
3. Add the hundreds. Ask: Do I need to regroup?
4. Add the thousands.

Studying the problem Read the problem and the notes beside it.

How much money did Denzel's school raise?

How much money did Marcus's school raise?

How much money is needed?

Denzel's school and his brother Marcus's school raised money to buy new sports equipment. Denzel's school raised $2,284. Marcus's school raised $2,479.

After both amounts are added, will there be enough money to purchase equipment priced at $4,761?

How can Denzel use a flowchart to solve the problem?

Studying the solution A **flowchart** is a graphic organizer that you can use to add numbers with four digits. Denzel used this flowchart to find out how much money the schools had all together.

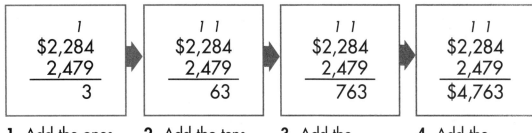

1. Add the ones. Ask: Do I need to regroup?
2. Add the tens. Ask: Do I need to regroup?
3. Add the hundreds. Ask: Do I need to regroup?
4. Add the thousands.

Denzel found that the schools had raised $4,763 all together. He knew that the schools had enough money to buy the equipment at $4,761.

Understanding the solution Read what Denzel wrote to explain how he used a flowchart to solve the problem.

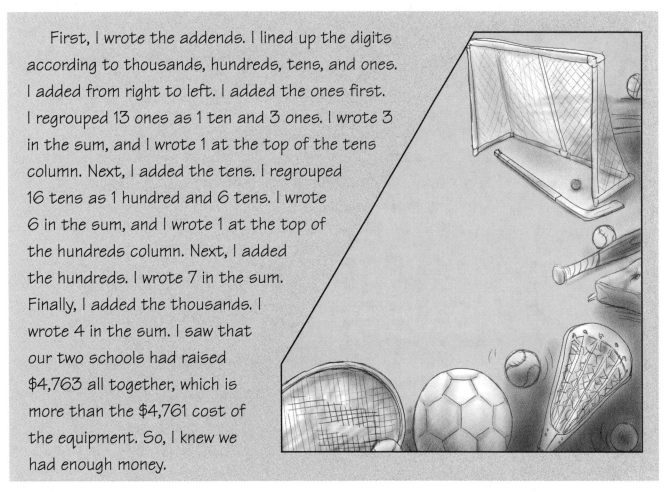

First, I wrote the addends. I lined up the digits according to thousands, hundreds, tens, and ones. I added from right to left. I added the ones first. I regrouped 13 ones as 1 ten and 3 ones. I wrote 3 in the sum, and I wrote 1 at the top of the tens column. Next, I added the tens. I regrouped 16 tens as 1 hundred and 6 tens. I wrote 6 in the sum, and I wrote 1 at the top of the hundreds column. Next, I added the hundreds. I wrote 7 in the sum. Finally, I added the thousands. I wrote 4 in the sum. I saw that our two schools had raised $4,763 all together, which is more than the $4,761 cost of the equipment. So, I knew we had enough money.

Solve a Problem

Studying the problem   Read the problem. As you read, think about how you could use a flowchart to solve the problem.

Shaka's third-grade class is studying the rivers of the United States. The Mississippi River, which is 2,340 miles long, is the second-longest river in the United States. The Missouri River, which is 2,540 miles long, is the longest. However, the Mississippi River is deeper and has more water than the Missouri River. The Red River is 1,290 miles long. Both the Missouri River and the Red River flow into the Mississippi River.

What is the combined length of the Mississippi River and the Red River?

Finding the solution   Use the information from the problem to complete this flowchart. Then write your answer below.

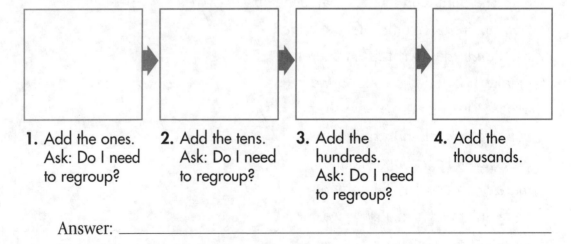

1. Add the ones. Ask: Do I need to regroup?
2. Add the tens. Ask: Do I need to regroup?
3. Add the hundreds. Ask: Do I need to regroup?
4. Add the thousands.

Answer: _____

Explaining the solution

Reread "Understanding the solution" on page 25, which tells how Denzel used a flowchart to solve a problem. Then write your own explanation of how you completed the flowchart on page 26 and found your solution.

_____

_____

_____

_____

_____

_____

_____

Applying the solution

Use your flowchart on page 26 to answer these questions.

1. Which addends did you use to solve the problem?
   _____

2. Did you have to regroup at any time? If so, when?
   _____

3. Which digits in the addends are in the hundreds place?
   _____

4. Use the combined length of the Mississippi and Red rivers and the length of the Missouri River to find the length of the three rivers all together. Write the problem and the solution.
   _____

5. The Arkansas River, which is 1,460 miles long, also flows into the Mississippi River. What is the combined length of the Arkansas and the Mississippi rivers?
   _____

Learn More About Addition

Thinking about the strategy

You may sometimes want to find the sum of three-digit numbers, but you may not need an exact sum. For example, you might want to know about how much two toy cars will cost. Perhaps you want to know about how many trading cards are in your collection.

You can use estimation to find the sum of three-digit numbers quickly. An estimated sum is close to the exact sum.

A **number line** is a graphic organizer that you can use to round three-digit numbers to the nearest hundred and to estimate their sums.

How can Luka find the estimated sum of 264 and 422? Luka used these number lines to find the answer.

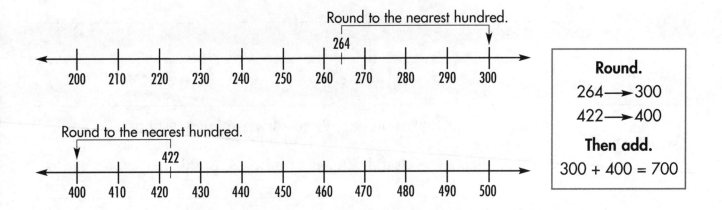

Answer: Luka found the estimated sum of 264 and 422 to be 700.

Understanding the solution

Read what Luka wrote to explain how he used number lines to solve the problem.

Find which hundreds the numbers come between.

Place the numbers on the number lines.

Round each number to the nearest hundred.

Add the rounded numbers.

To make a number line for 264, I marked off tens between 200 and 300. I placed the number 264 on the number line. I could see that 264 is closer to 300 than to 200. I rounded 264 to 300. To make a number line for 422, I marked off tens from 400 to 500. I placed 422 on the number line and could see that 422 is closer to 400 than to 500. I rounded 422 to 400. Then I added 300 and 400 and got the estimated sum of 700.

Solve a Problem

Finding the solution

Alina and her sister collect stamps from all over the world. In one album, they have 539 stamps. In another album, they have 371 stamps. About how many stamps do Alina and her sister have?

Complete these number lines to find about how many stamps Alina and her sister have. Then write your answer below.

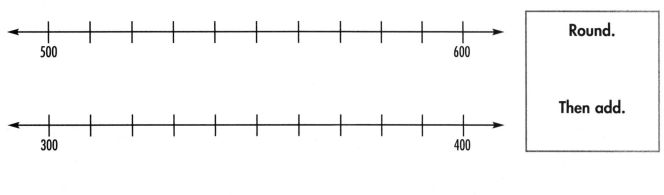

Round.

Then add.

Answer: _____

Explaining the solution

Write an explanation of how you found the answer by completing the number lines above.

_____

_____

_____

_____

_____

_____

_____

_____

Numbers in Context

Read "The Playground." Think about the ways that numbers are used in the selection. Then answer items A–C on page 31.

The Playground

Amos is learning about the United States Constitution. The Constitution was written to guide the nation's leaders and to protect people's rights. Among those rights is the right to petition. A petition is something that people sign to show that they are for or against an idea or a cause.

When Amos's teacher, Mr. Douglas, asked if anyone in the class had ever signed a petition, Amos raised his hand. "Please explain," Mr. Douglas said.

"Well," Amos said, "I myself have never signed a petition, but my parents have. There's a playground on the corner of our street. For a long time, people dumped their trash there. My parents asked the city to clean it up many times, but no one ever showed up to do the job. So my parents and some other grown-ups in the neighborhood had a meeting. They wrote a petition asking the mayor to clean up the playground right away."

"Then what happened?" Mr. Douglas asked.

"Then they made copies of the petition," Amos said. "They took it around to just about every apartment and house in the neighborhood. My mom collected 417 signatures. My dad collected 385 signatures. Several other people collected signatures too. All together, 2,759 people signed the petition. Another 1,482 people said they would like to see the playground cleaned up, but they didn't want to sign anything.

"My father brought the petition to city hall," Amos said. "When the mayor saw how many people had signed it, he agreed to clean up the playground right away. Once all the trash was gone, the neighbors all pitched in to paint and fix the playground equipment. They even planted flowers and some small trees. It's a great place to play now."

A. Including those people who signed the petition and those who didn't, how many people agreed that the playground should be cleaned up? Use the information from page 30 to complete this flowchart. Then write your answer below.

1. Add the ones. Ask: Do I need to regroup?
2. Add the tens. Ask: Do I need to regroup?
3. Add the hundreds. Ask: Do I need to regroup?
4. Add the thousands.

Answer: _____

B. About how many signatures did Amos's mom and dad collect together? Use the information from page 30 to complete these number lines. Then write your answer below.

Round.

Then add.

Answer: _____

C. Explain your solution to either item A or item B above.

_____
_____
_____
_____
_____

Check Your Understanding

Fill in the letter of the correct answers to questions 1–8.
Write your answers to questions 9 and 10.

1. The Jenkins School received bills for repairs done last summer. One bill was for $5,765. The other bill was for $3,896. How much did the school have to pay for the repairs?
   - Ⓐ $9,551
   - Ⓑ $8,651
   - Ⓒ $9,661
   - Ⓓ $8,561

2. Wakeland has two daily newspapers. The Morning News is delivered to 4,148 homes. The Sun Reporter is delivered to 2,961 homes. How many newspapers are delivered every day?
   - Ⓐ 7,109 newspapers
   - Ⓑ 6,119 newspapers
   - Ⓒ 7,019 newspapers
   - Ⓓ 6,009 newspapers

3. The lunchroom floor in Josie's school is tiled with 1,899 tiles. The lunchroom in Walter's school has 1,421 tiles. How many tiles are in both lunchrooms?
   - Ⓐ 2,211 tiles
   - Ⓑ 2,310 tiles
   - Ⓒ 3,200 tiles
   - Ⓓ 3,320 tiles

4. A large video store sold 1,319 copies of a movie on one day and 1,755 copies the next day. How many copies of the film did the store sell in two days?
   - Ⓐ 2,165 copies
   - Ⓑ 3,074 copies
   - Ⓒ 3,164 copies
   - Ⓓ 2,072 copies

5. Two schools entered all their third-grade students in a math contest. One school has 335 third graders. The other school has 287 third graders. About how many students will be in the contest?
   - Ⓐ 500 students
   - Ⓑ 400 students
   - Ⓒ 700 students
   - Ⓓ 600 students

6. Ms. Kim has $183 in one bank account and $215 in another account. About how much money does Ms. Kim have in the two accounts?
   - Ⓐ $200
   - Ⓑ $300
   - Ⓒ $400
   - Ⓓ $500

7. Pedro's mom is using circles and triangle shapes to make a quilt. She needs 462 circle pieces and 424 triangle pieces for the pattern. About how many pieces does she need all together?
   - Ⓐ 900 pieces
   - Ⓑ 800 pieces
   - Ⓒ 700 pieces
   - Ⓓ 1,000 pieces

8. Paper supplies for Hank's school cost $393 last month. Paper supplies for Paul's school cost $359. About how much did paper supplies for both schools cost all together?
   - Ⓐ $600
   - Ⓑ $700
   - Ⓒ $800
   - Ⓓ $900

9. Last fall, workers planted 6,275 tulip bulbs and 2,890 daffodil bulbs around the pond in City Park. How many bulbs did they plant all together?

10. At a used-book store, Mr. Hubble bought a set of the eleventh edition of the *Encyclopedia Britannica* for $275 and a set of 1920's children's books for $145. To the nearest hundred dollars, about how much did Mr. Hubble spend? Explain how you found your answer.

Extend Your Learning

- *Book Talk*

    Talk to someone at a bookstore or use the Internet to find the prices of two computer games, each over $100. Then estimate, to the nearest hundred dollars, about how much the games would cost together.

- *Social Studies: Down the River*

    Work with a partner to find the lengths of the two longest rivers in your state. Figure out how far you would travel if you traveled the lengths of both rivers.

STRATEGY FOUR: Applying Subtraction

Learn About Subtraction

Thinking about the strategy

You use subtraction to find the difference between two numbers. You sometimes have to subtract large numbers. A four-digit number has a digit in the ones place, tens place, hundreds place and thousands place. To subtract numbers with four digits, line up the digits by place value. Then subtract from right to left. Regroup when a digit in the number you are subtracting from is less than a digit in the number you are subtracting.

A flowchart can help you subtract numbers with four digits.

What is 2,334 subtracted from 5,316? As you subtract each column, you may need to regroup.

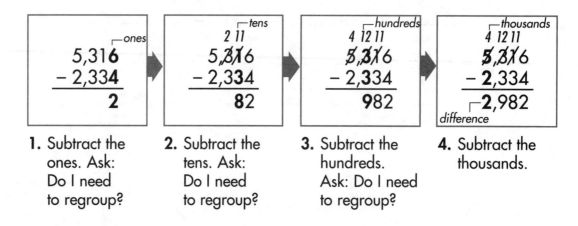

1. Subtract the ones. Ask: Do I need to regroup?
2. Subtract the tens. Ask: Do I need to regroup?
3. Subtract the hundreds. Ask: Do I need to regroup?
4. Subtract the thousands.

Studying the problem   Read the problem and the notes beside it.

How much was sold two years ago?

How much was sold last year?

LuAnn's parents work in a hardware store. Two years ago, they sold 6,245 feet of wiring. Last year, they sold 4,657 feet of wiring.

How much more wiring was sold two years ago than last year?

How can LuAnn use a flowchart to solve the problem?

34

Studying the solution   A **flowchart** is a graphic organizer that you can use to subtract numbers with four digits. LuAnn used this flowchart to find how much more wiring was sold two years ago than last year.

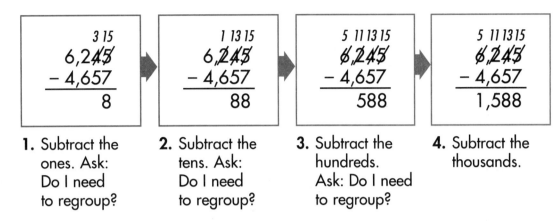

1. Subtract the ones. Ask: Do I need to regroup?
2. Subtract the tens. Ask: Do I need to regroup?
3. Subtract the hundreds. Ask: Do I need to regroup?
4. Subtract the thousands.

LuAnn found that 1,588 feet more wiring was sold two years ago than last year.

Understanding the solution   Read what LuAnn wrote to explain how she used a flowchart to solve the problem.

First, I saw that the larger number is 6,245. So, I knew that I had to subtract the smaller number, 4,657 from 6,245. I lined up the digits according to thousands, hundreds, tens, and ones. Then I subtracted, moving from right to left. When the number I was subtracting was greater than the number I was subtracting from, I regrouped. I regrouped 1 ten as 10 ones. I also had to regroup 1 hundred as 10 tens, and 1 thousand as 10 hundreds. I could see that 1,588 feet more wiring was sold two years ago than last year.

Solve a Problem

Studying the problem Read the problem. As you read, think about how you could use a flowchart to solve the problem.

Enrique's class watched a film about the famous journey of two explorers named Lewis and Clark.

In May 1804, Lewis and Clark set off from the mouth of the Missouri River in search of the Pacific Ocean. Clark kept a daily journal during the journey west. In it, he noted that by the time they reached the Pacific Ocean, in November 1805, they had traveled 4,162 miles by river and land.

Enrique and his group mapped out Lewis and Clark's journey. They figured out that, because of today's roads, bridges, and tunnels, the distance from the mouth of the Missouri River to the Pacific Ocean is about 2,165 miles.

What is the difference in length between the journey then and now?

Finding the solution Use the information from the problem to complete this flowchart. Then write your answer below.

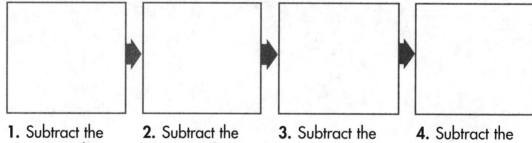

1. Subtract the ones. Ask: Do I need to regroup?
2. Subtract the tens. Ask: Do I need to regroup?
3. Subtract the hundreds. Ask: Do I need to regroup?
4. Subtract the thousands.

Answer: _____

Explaining the solution

Reread "Understanding the solution" on page 35, which tells how LuAnn used a flowchart to solve a problem. Then write your own explanation of how you completed the flowchart on page 36 and found your solution.

_____

_____

_____

_____

_____

_____

_____

Applying the solution

Use your flowchart on page 36 to answer these questions.

1. How many thousands are in the distance that Lewis and Clark traveled? _____

2. Did you have to regroup to subtract any digits? If so, when?

   _____

3. After you regrouped the ones column, how many ones were there?

   _____

4. Maria's group mapped a different route and figured that today, Lewis and Clark's trip would cover 2,341 miles. How much shorter is this route than the one taken by Lewis and Clark?

   _____

5. What's the difference between the mileage that Enrique's group used and the mileage that Maria's group used?

   _____

Learn More About Subtraction

Thinking about the strategy

When you do not need to know the exact difference between numbers, you can use estimation. An estimated difference is close to the exact difference.

A **number line** is a graphic organizer that you can use to round three-digit numbers to the nearest hundred and to estimate their differences.

How can Sam find the estimated difference between 416 and 173? Sam used these number lines to find the answer.

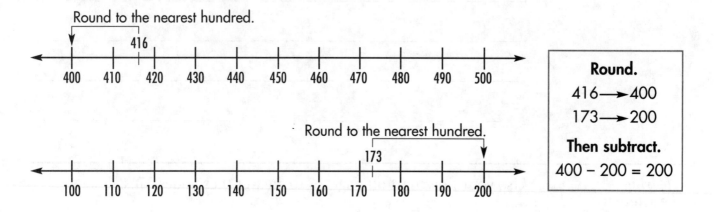

Answer: Sam found the estimated difference between 416 and 173 to be 200.

Understanding the solution

Read what Sam wrote to explain how he used number lines to solve the problem.

Place the numbers on the number lines.

Round up or down to the nearest hundred.

Subtract the rounded numbers.

To make a number line for 416, I marked off tens between 400 and 500. I placed the number 416 on the number line. I could see that 416 is closer to 400 than to 500. I rounded 416 to 400. To make a number line for 173, I marked off tens between 100 and 200. I placed the number 173 on the number line. I could see that 173 is closer to 200 than to 100. I rounded 173 to 200. Then I subtracted 200 from 400 and got the estimated difference of 200.

Solve a Problem

Finding the solution

Will lives in New York City. His grandparents live in Atlanta, Georgia, 867 miles away. His aunt lives 632 miles away from New York City, in Charlotte, North Carolina. About how much closer is New York City to Will's aunt's home than it is to his grandparents' home?

Complete these number lines to find about how much closer it is to Will's aunt's home than it is to his grandparents' home. Then write your answer below.

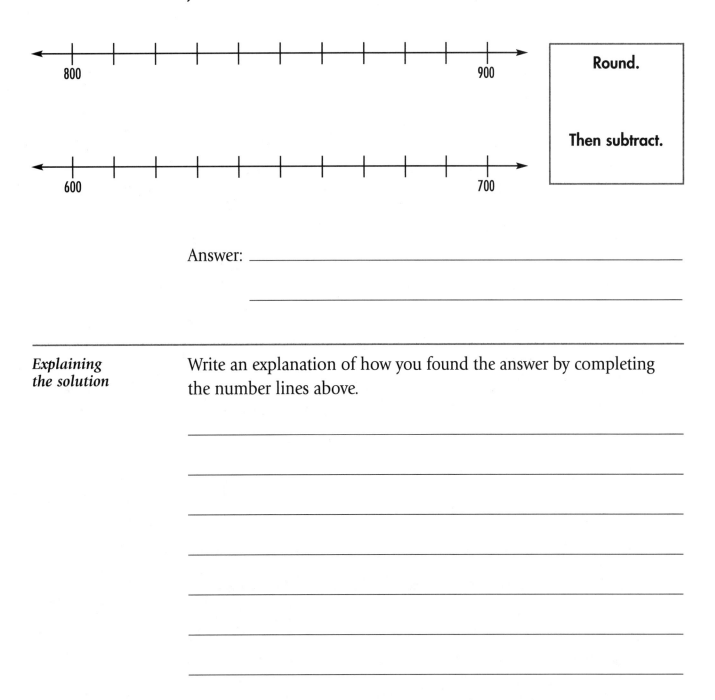

Answer: _____

Explaining the solution

Write an explanation of how you found the answer by completing the number lines above.

_____

_____

_____

_____

_____

_____

Numbers in Context

Read "Visiting the New Zoo." Think about the ways that numbers are used in the selection. Then answer items A–C on page 41.

Visiting the New Zoo

The Eastbrook Zoo closed 18 months ago. Last weekend, it reopened its gates under a new name, the Eastbrook Nature Park and Zoo.

In a television interview, the head zookeeper, Mrs. Barry, reported that opening weekend was quite busy. On Saturday, 2,538 men, women, and children visited the zoo. On Sunday, 3,219 visitors came through the gates.

"Obviously, people have been looking forward to this weekend," Mrs. Barry said. Mrs. Barry went on to describe some of the changes that were made during the time the zoo was closed. Now the big cats, including the tigers, lions, and panthers, are housed in an area that looks like a real jungle. There are trees, rocky ledges, waterfalls, and a flowing river.

A new area was also created for monkeys, with plenty of trees for them to swing from branch to branch.

"The animals seem much happier in their new, natural surroundings," Mrs. Barry said. Besides updating the animal houses, some new animals were added. "When the zoo closed last year, we had 275 animals of about 50 species," Mrs. Barry said. "Now we have 368 animals of 65 species."

The Eastbrook Nature Park and Zoo is located on Route 111, five miles north of the Applewood Mall. The zoo is open Monday through Friday from 10:00 A.M. to 5:00 P.M. On Saturdays and Sundays, it is open from 8:00 A.M. to 5:00 P.M. Tickets are $3.50 for children ages 5 to 12 and $6.00 for adults.

A. How many more visitors did the zoo have on Sunday than it had on Saturday? Use the information from page 40 to complete this flowchart. Then write your answer below.

1. Subtract the ones. Ask: Do I need to regroup?
2. Subtract the tens. Ask: Do I need to regroup?
3. Subtract the hundreds. Ask: Do I need to regroup?
4. Subtract the thousands.

Answer: _____

B. About how many more animals does the new zoo have than the old zoo had? Use the information from page 40 to complete these number lines. Then write your answer below.

Round.

Then subtract.

Answer: _____

C. Explain your solution to either item A or item B above.

_____

_____

_____

_____

_____

41

Check Your Understanding

Fill in the letter of the correct answers to questions 1–8.
Write your answers to questions 9 and 10.

1. Mercury is about 3,032 miles wide. Mars is about 4,222 miles wide. How much wider is Mars than Mercury?
   Ⓐ 1,290 miles
   Ⓑ 1,354 miles
   Ⓒ 1,190 miles
   Ⓓ 892 miles

2. Paco read that there are about 5,940 kinds of living reptiles. Of those, 3,025 are lizards, crocodiles, and alligators. The rest are turtles and snakes. How many kinds of turtles and snakes are there?
   Ⓐ 2,915 kinds
   Ⓑ 2,825 kinds
   Ⓒ 1,925 kinds
   Ⓓ 1,815 kinds

3. The town of Fosbeck had set aside $6,129 for new recycling bins. However, the new bins cost only $2,335. How much money will the town have set aside after the bins are purchased?
   Ⓐ $4,884
   Ⓑ $4,214
   Ⓒ $3,854
   Ⓓ $3,794

4. Keesha's sister bought a used car for $8,500. She used her savings of $5,632 and borrowed the rest from her dad. How much did Keesha's sister borrow?
   Ⓐ $3,978
   Ⓑ $2,868
   Ⓒ $3,872
   Ⓓ $2,862

5. Apex Company sells the Bakey stove for $788. Barrera's sells the same stove for $537. What is the estimated difference between the two prices?
   Ⓐ $300
   Ⓑ $200
   Ⓒ $100
   Ⓓ $600

6. Neighborhood yard sales raised $904 in the spring and $532 in the fall. To the nearest hundred dollars, about how much less money did neighbors make at the fall sale?
   Ⓐ $300   Ⓒ $500
   Ⓑ $460   Ⓓ $400

7. Word entries take up 779 pages of a dictionary with 857 pages. To the nearest hundred, about how many pages show other information?
   Ⓐ 100 pages
   Ⓑ 150 pages
   Ⓒ 200 pages
   Ⓓ 250 pages

8. Stella's class used paper flowers to make an American flag. They used 444 red flowers, 421 white flowers, and 128 blue flowers. To the nearest hundred, about how many more red flowers than blue flowers did they use?
   Ⓐ 100 flowers
   Ⓑ 200 flowers
   Ⓒ 300 flowers
   Ⓓ 400 flowers

9. An's family went to two stores to shop for a new computer. At the first store, the computer they wanted cost $1,299. At the second store, the same computer cost $1,650. How much money will they save if they buy the computer at the first store?

_____

10. Workers are putting up a new apartment building next door to Jean's building. Jean's building is 126 feet high. The new building will be 208 feet high. To the nearest hundred, what is the difference in the heights of the two buildings. Explain how you found your answer.

_____
_____
_____
_____
_____
_____
_____
_____

Extend Your Learning

- *Travel Plans*

    In a group, plan a trip in which you travel across the country and then back home. Plan a different route for going and for coming back. Figure out about how many miles the trip will be each way. Then find what part of your trip is longer and by how many miles.

- *Social Studies: World's Tallest Buildings*

    Work with a partner to find the height, in feet and in meters, of five of the world's tallest buildings. Estimate the difference in the heights of two of the buildings.

STRATEGY FIVE: Applying Multiplication

Learn About Multiplication

Thinking about the strategy

In multiplication, the numbers that you multiply are called factors. Your answer is the product. When the number 1 is one of the factors, the product is always the same as the other factor.

There are 3 pencils in 1 drawer. How many pencils are in 1 drawer?
$1 \times 3 = \boxed{?}$

 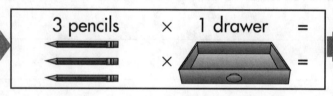 

When 0 is one of the factors, the product is always 0. There are 0 pencils in 4 drawers. How many pencils are in the 4 drawers? $0 \times 4 = \boxed{?}$

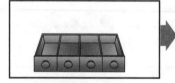 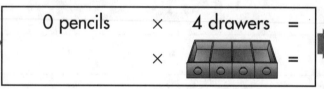

Studying the problem Read the problem and the notes beside it.

How many packs of each color are there?

How many boxes of each color are there?

What factors could Eve use to find the number of packs of each color of paper?

> Mr. Owen asked Eve to count and list the packs of construction paper in the supply closet. Eve counted 7 packs of paper in the box labeled "red." She counted 8 packs of paper in the box labeled "yellow." There were 5 boxes labeled "black," but all of them were empty. The 1 box labeled "green" was also empty.
>
> How many packs of each color were in the supply closet?

How can Eve use a multiplication table to solve the problem?

Studying the solution

A **multiplication table** is a graphic organizer that you can use to find a product when one of the factors is 1 or when one of the factors is 0. After Eve counted the packs of paper in the supply closet, Mr. Owen asked her to use multiplication to show the total number of packs of each color. Eve used the numbers from her list to make this multiplication table and to show the totals for each color.

Any Number	×	1	=	The Number
7	×	1	=	7
8	×	1	=	8

Number	×	0	=	0
5	×	0	=	0
1	×	0	=	0

Eve found that there were 7 packs of red paper ($7 \times 1 = 7$) and 8 packs of yellow paper ($8 \times 1 = 8$). She found that there were 0 packs of black paper ($5 \times 0 = 0$) and 0 packs of green paper ($1 \times 0 = 0$).

Understanding the solution

Read what Eve wrote to explain how she used a multiplication table to solve the problem.

To use multiplication to show the total number of packs of red paper, I multiplied 7 (packs of red paper) × 1 (number of boxes). Since the product of any number times 1 equals the number, I found that 7 × 1 = 7. Then I multiplied 8 (packs of yellow paper) × 1 (number of boxes) and got 8, to show there were 8 packs of yellow paper.

To use multiplication to show the total number of packs of black paper, I multiplied 5 (number of boxes) × 0 (number of packs of black paper). Since the product of any number times 0 equals 0, I found that 5 × 0 = 0. Then I multiplied 1 × 0 and got 0, to show there were 0 packs of green paper.

Solve a Problem

Studying the problem Read the problem. As you read, think about how you could use a multiplication table to solve the problem.

For a science project, José used the Internet to find the monthly rainfall totals for his region. He found that last year, 1 inch of rain had fallen during each of the 3 months of spring. This year, no rain had fallen during the 3 months of spring.

Based on the information José found, how many inches of rain fell last spring? How many inches of rain fell this spring?

Finding the solution Use the information from the problem to complete this multiplication table. Then write your answer below.

Any Number	×	1	=	The Number
	×	1	=	

Number	×	0	=	0
	×	0	=	

Answer: _____

_____

46

Explaining the solution

Reread "Understanding the solution" on page 45, which tells how Eve used a multiplication table to solve a problem. Then write your own explanation of how you completed the multiplication table on page 46 and found your solution.

_____

_____

_____

_____

_____

_____

_____

Applying the solution

Use your multiplication table on page 46 to answer these questions.

1. What two factors did you use to find how many inches of rain fell last spring? _____

2. What two factors did you use to find how many inches of rain fell this spring? _____

3. What two factors would you use to find how much rain fell the first 2 months of summer if 3 inches fell each month? _____

4. What two factors would you use to find how much rain fell the first 5 months of last year if no rain fell at all? _____

5. José learned that last year, 1 inch of rain fell for each of the 4 weeks in September. What numbers should you use as factors to find how many inches of rain fell in September?

47

Learn More About Multiplication

Thinking about the strategy

You may sometimes have to multiply three numbers. You can multiply three numbers in any order. Usually, it's easier to multiply numbers in the order that they are given in the problem. However, you may wish to change the order of the factors. Changing the order of the factors will not change the answer. Just be sure to multiply all the factors.

A **step diagram** is a graphic organizer that you can use to multiply three factors.

How can Janet find the product of $2 \times 6 \times 4$? Janet used this step diagram to find the answer.

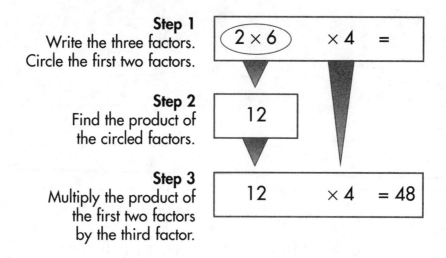

Step 1 Write the three factors. Circle the first two factors.

Step 2 Find the product of the circled factors.

Step 3 Multiply the product of the first two factors by the third factor.

Answer: *Janet found that the product of $2 \times 6 \times 4$ is 48.*

Understanding the solution

Read what Janet wrote to explain how she used a step diagram to solve the problem.

Multiply two of the factors.

Then multiply the product times the third factor.

To check your answer, multiply the factors in a different order.

First, I wrote the three factors and circled the 2 and 6 to multiply first. Next, I multiplied 2×6 and wrote the product, 12. Last, I multiplied the product of the first two factors, 12, times the third factor, 4. I found that $12 \times 4 = 48$. I also could have found the answer by multiplying $6 \times 4 = 24$ and $24 \times 2 = 48$; or $2 \times 4 = 8$ and $8 \times 6 = 48$.

Solve a Problem

Finding the solution

Chandra drew a design for a small garden. The design showed 5 rows of plants. Each row would have 5 vegetable plants. Chandra is planting tomatoes, green beans, peppers, corn, and broccoli. Chandra's mother decided to plant a garden with yellow beans, summer squash, pumpkins, zucchini, and eggplant. If she follows Chandra's garden design, how many plants will Chandra and her mother have all together?

Complete this step diagram to find how many plants Chandra and her mother planned to have in their gardens. Then write your answer below.

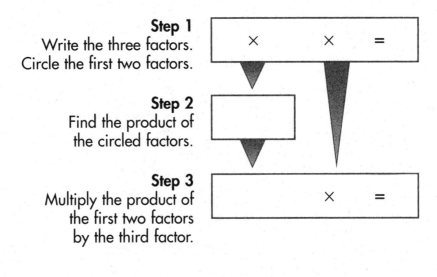

Step 1
Write the three factors.
Circle the first two factors.

Step 2
Find the product of the circled factors.

Step 3
Multiply the product of the first two factors by the third factor.

Answer: _____

Explaining the solution

Write an explanation of how you found the answer by completing the step diagram above.

_____
_____
_____
_____
_____
_____

Numbers in Context

Read "A Trip to the Aquarium." Think about the ways that numbers are used in the selection. Then answer items A–C on page 51.

A Trip to the Aquarium

Ms. Andrews, Mike's teacher, spoke to both Mike's and another class before the students climbed on the buses. "Once we get to the aquarium," she said, "each class will separate into 7 groups of 3 students and an adult leader. Be sure that you stay with your group at all times."

"I can't wait to see the sharks," Nick said to Mike as they got on the bus.

Nick, Mike, and Anna were in the same group. "Can we go see the sharks first?" Nick asked their leader, Mrs. Chin, as soon as they arrived.

Mrs. Chin was Mike's mother and a scientist. She knew all about sharks and other sea creatures. As they walked toward the sharks, Mrs. Chin said, "There are about 400 different kinds of sharks. Most are not dangerous to people. In fact, you're more likely to get bitten by a squirrel than a shark!

"Of course, you have to use some sense," Mrs. Chin said. "Don't swim at dawn, dusk, or at night. That's when sharks swim into shallow water to find food. And don't swim if you have a cut that's bleeding. Sharks have a great sense of smell. They can smell blood for miles away."

The aquarium had 1 huge shark tank and 6 smaller shark tanks. A sign on the huge tank said that it held a Great White and 8 other sharks. All of the 6 smaller tanks were being repaired, so these tanks were empty.

Nick, Mike, and Anna got as close to the huge tank as they could. At that moment, the Great White Shark swam by with its mouth wide open.

"Look at those teeth!" Mike and Anna said at the same time.

"I think I'll take my chances with a squirrel," Mike said.

A. How many sharks were in the huge tank? How many sharks were in the smaller tanks? Use the information from page 50 to complete this multiplication table. Then write your answer below.

Any Number	×	1	=	The Number		Number	×	0	=	0
	×	1	=				×	0	=	
	×	1	=				×	0	=	

Answer: _____

B. How many students went on the field trip to the aquarium? Use the information from page 50 to complete this step diagram. Then write your answer below.

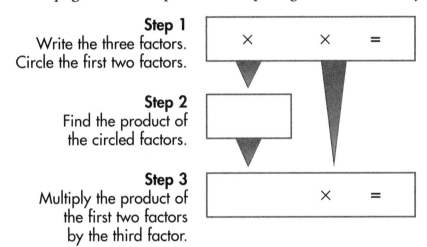

Step 1
Write the three factors.
Circle the first two factors.

Step 2
Find the product of the circled factors.

Step 3
Multiply the product of the first two factors by the third factor.

Answer: _____

C. Explain your solution to either item A or item B above.

_____

_____

_____

_____

_____

_____

Check Your Understanding

Fill in the letter of the correct answers to questions 1–8.
Write your answers to questions 9 and 10.

1. Wally's mom dug 7 holes in the ground and planted 1 tulip bulb in each hole. How many tulip bulbs did she plant?
   - Ⓐ 14 tulip bulbs
   - Ⓑ 8 tulip bulbs
   - Ⓒ 0 tulip bulbs
   - Ⓓ 7 tulip bulbs

2. Kim's dad built 2 bookcases with 5 shelves in each bookcase. Kim has not put any books in the bookcase yet. How many shelves are filled with books?
   - Ⓐ 0 shelves
   - Ⓑ 5 shelves
   - Ⓒ 10 shelves
   - Ⓓ 2 shelves

3. Mr. Adams made 4 pizzas and cut each pizza into 8 pieces. Jim and his friends ate 1 whole pizza. How many pieces did they eat?
   - Ⓐ 4 pieces
   - Ⓑ 8 pieces
   - Ⓒ 12 pieces
   - Ⓓ 16 pieces

4. Erin's dad works at the Berson Civic Center. He was able to get Erin and 3 of her friends $10 tickets to the ice show for free. How much did Erin's dad pay for the 4 tickets?
   - Ⓐ $40.00
   - Ⓑ $30.00
   - Ⓒ $10.00
   - Ⓓ $0.00

5. For two days last week, Gloria's dad parked his car in a city lot that charges $1.00 per hour. The car was parked for 9 hours both days. How much did he pay in all?
   - Ⓐ $10.00
   - Ⓑ $18.00
   - Ⓒ $9.00
   - Ⓓ $20.00

6. Milo helped the PTA get ready for their yearly party. He set up 3 rows of 3 tables each in the gym. He put 6 chairs at each table for the guests. How many guests does the PTA expect at the party?
   - Ⓐ 12 guests
   - Ⓑ 18 guests
   - Ⓒ 54 guests
   - Ⓓ 36 guests

7. Raina figured out that she could pour 8 large glasses of milk from 1 carton. How many glasses could she pour from 3 cartons of milk?
   - Ⓐ 12 glasses
   - Ⓑ 32 glasses
   - Ⓒ 16 glasses
   - Ⓓ 24 glasses

8. Annie's mom has a shoe rack, which is empty now. The shoe rack has 6 rows of plastic pockets for shoes, with 4 pockets in each row. How many pairs of shoes are in the shoe rack?
   - Ⓐ 0 pairs of shoes
   - Ⓑ 24 pairs of shoes
   - Ⓒ 12 pairs of shoes
   - Ⓓ 48 pairs of shoes

9. To make yogurt sundaes, Vanessa took 8 large glass bowls from the cabinet. She put 1 cup of yogurt into each bowl. She put 1 cup of strawberries into each of 6 bowls. She put 1 cup of blueberries into each of the other 2 bowls. How many cups of yogurt, strawberries, and blueberries did Vanessa use to make the sundaes?

10. Pete has a collection of wild-animal stickers. His sticker book has 24 pages, but only 4 pages have stickers so far. Each of those pages has 3 wild-animal stickers on it. Next to each sticker, Pete writes 2 facts about the animal pictured on the sticker. So far, how many animal facts has Pete written in his sticker book? Explain how you found your answer.

Extend Your Learning

- *How Many?*

    Use multiplication to estimate the number of students in your school. Count the students in your classroom. Multiply by the number of classrooms in the same grade. Multiply the product by the number of grades in your school. Compare your answer to the actual number of students in your school.

- *Science: Animal Math*

    With a partner, use the Internet to find out more about sharks or another sea creature of your choice. List facts about size, eating habits, speed, and so on. Use the facts to make up multiplication problems that use the factors 1 and 0.

STRATEGY SIX: Applying Division

Learn About Division

Thinking about the strategy

When you divide, you separate a number into equal amounts. In division, the number you divide is the dividend. The number you divide by is the divisor. The result is the quotient.

$$\text{divisor} \rightarrow 4\overline{)8} \leftarrow \text{dividend}$$
$$\phantom{\text{divisor} \rightarrow 4\overline{)}}\ 2 \leftarrow \text{quotient}$$

You can use division to divide two-digit numbers by one-digit numbers. When a number does not divide evenly, the amount left over is called the remainder. The remainder is part of the quotient. The remainder is never greater than the divisor.

You can use a flowchart to divide 79 by 3. What steps should you follow? Why is R1 written in the quotient?

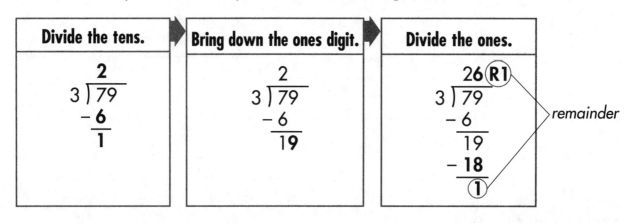

Studying the problem Read the problem and the notes beside it.

How many new cards does Cal have?

How many cards does each page hold?

What operation should Cal use?

For his birthday, Cal received 55 new baseball cards for his collection. To protect the cards, Cal keeps them in plastic pages in a three-ring binder. Each page holds 4 cards. Cal is all out of plastic pages.

What is the fewest number of plastic pages that Cal needs for storing his new cards?

How can Cal use a flowchart to solve the problem?

Studying the solution

A **flowchart** is a graphic organizer that you can use to find how many equal parts are in a number. Cal used this flowchart to find how many groups of 4 cards were in the 55 cards he received. Then he knew how many plastic pages he needed.

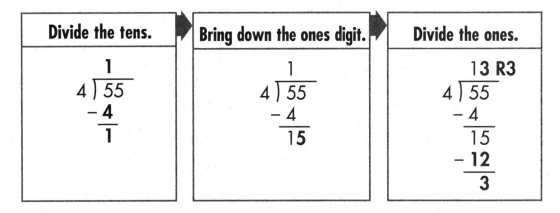

Cal found that he needed 14 new pages. He would put 4 cards on each of the first 13 pages. He would put the remaining 3 cards on the 14th page.

Understanding the solution

Read what Cal wrote to explain how he used a flowchart to solve the problem.

> I first had to find how many 4s are in 55. I wrote the problem, making 55 the dividend and 4 the divisor. First, I divided the tens. I know that 4 goes into 5 one time, so I wrote 1 above the 5 in the tens place. Then I multiplied 1 times 4 and got 4. I wrote 4 in the tens place under the 5. I subtracted and got 1. I brought down the ones digit, which was 5. I had a total of 15 ones. I divided the ones. I divided 4 into 15. I know that $4 \times 4 = 16$. Since 16 is 1 more than 15, I knew that 4 goes into 15 fewer than 4 times. I tried 3. I wrote 3 in the ones place in the quotient. I multiplied 3 times 4 and got 12. I wrote 12 under the 15 ones. I subtracted 12 from 15 and got 3. I showed the remainder 3 as R3 in the quotient. I have too many cards to fit into 13 pages, so I will have to put 3 cards on page 14.

Solve a Problem

Studying the problem   Read the problem. As you read, think about how you could use a flowchart to solve the problem.

> Every September, Martha's mom organizes a clothing drive in her neighborhood. She asks people to donate heavy coats and jackets that children have outgrown but that are still in good condition. This year, Martha's mom collected 93 jackets, coats, and snowsuits in a variety of sizes. Most of them looked brand-new.
>
> All the donated items were cleaned. Then Martha helped her mom pack the jackets, coats, and snowsuits into 8 large boxes. They divided the items equally among the boxes and shipped the boxes to a children's charity in the city. The items that were left over were shipped in another box, along with hats and mittens.
>
> How many items did they put into each large box? How many items were left over and shipped with the hats and mittens?

Finding the solution   Use the information from the problem to complete this flowchart. Then write your answer below.

Divide the tens.	**Bring down the ones digit.**	**Divide the ones.**

Answer: _____

Explaining the solution

Reread "Understanding the solution" on page 55, which tells how Cal used a flowchart to solve a problem. Then write your own explanation of how you completed the flowchart on page 56 and found your solution.

_____

_____

_____

_____

_____

_____

_____

_____

Applying the solution

Use your flowchart on page 56 to answer these questions.

1. What number was the divisor in the problem? _____

2. What number was the dividend? _____

3. Did the divisor go into the dividend evenly? If your answer is no, how did you show the remainder in the quotient?

   _____

4. Martha's mom used almost one complete roll of packing tape to wrap 6 boxes. The roll contained 64 feet of tape. Martha's mom used the same amount of tape on each of the 6 boxes. The amount was in whole feet. How much tape did she use on each box?

   _____

5. How many feet of tape were left over?

   _____

Learn More About Division

Thinking about the strategy

Multiplication is the opposite of division. You can use multiplication facts to solve division problems.

> The product of the multiplication problem becomes the dividend in each division problem. One factor becomes the divisor. The other factor becomes the quotient.

factors — $3 \times 4 = 12$ — product

dividend — $12 \div 3 = 4$ — divisor, quotient

dividend — $12 \div 4 = 3$ — divisor, quotient

A **multiplication table** is a graphic organizer that you can use to solve division problems.

How can Ginny find how many times 6 goes into 18? What multiplication and division problems can she use to find the solution? Ginny used this multiplication table to find the answer.

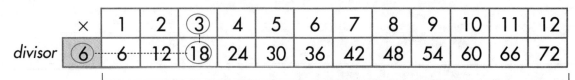

×	1	2	③	4	5	6	7	8	9	10	11	12
⑥	6	12	⑱	24	30	36	42	48	54	60	66	72

divisor

Ginny wrote each multiple of the divisor

Answer: Ginny found that 6 goes into 18, 3 times. To find the solution, she used 6 × 3 = 18 and 18 ÷ 6 = 3.

Understanding the solution

Read what Ginny wrote to explain how she used a multiplication table to solve the problem.

> Another way to find multiples is to skip count. For this problem, skip count by 6, the divisor.
>
> When you use a multiplication fact to solve a division problem, the factors become the divisor and the quotient.

I used multiplication facts for 6, the divisor. I wrote 6 in the first box of the table and circled it. Then I multiplied and wrote multiples of 6 across the table. Since I wanted to find how many times 6 goes into 18, I moved across the table to 18 and circled the 18. I saw that 18 was a multiple of 6 and 3. I circled the 3. I know that multiplication is the opposite of division. So, if 6 × 3 = 18, then 18 ÷ 6 = 3.

Solve a Problem

Finding the solution

Jason and a group of friends went on a hike with Jason's older brothers. The hike was a total of 10 miles. The group planned to stop every 2 miles to rest. How many stops, including the last stop at the end, did the group make on the hike?

Complete this multiplication table to figure out how many rest stops the group made. Then write your answer below. Write the multiplication and division facts you used to find the solution.

×	1	2	3	4	5	6	7	8	9	10	11	12
divisor												

Write each multiple of the divisor.

Answer: _____

Explaining the solution

Write an explanation of how you found the answer by completing the multiplication table above.

_____

_____

_____

_____

_____

_____

_____

_____

Numbers in Context

Read "Moving Art." Think about the ways that numbers are used in the selection. Then answer items A–C on page 61.

Moving Art

"What does the word *transportation* mean?" Mrs. Perez asked.

"I think it means how people get from one place to another," Damont answered, "like in a plane, train, or bus."

"That's right," Mrs. Perez said. "Now, can anyone tell me what the earliest form of transportation was? Do you know, Alice?"

Alice thought for a minute. "Was it the stagecoach?"

Mrs. Perez smiled. "Horse-drawn wagons were an early form of transportation, but how do you think the very first humans traveled?"

"On foot," Alice said.

"Sure," Mrs. Perez said. "It's actually still a good way for people to get around. Today, how people travel often depends on where they live. For example, some people who live in the desert travel by camel. Some people who live in the rain forest travel from place to place by canoe, on a river. In China, many people travel along crowded streets by bicycle."

Mrs. Perez and her 24 students discussed transportation. Then Mrs. Perez brought out a roll of paper. She said, "We're going to paint a mural. First, split up into groups of 4 students. Then count how many groups there are. Then divide this roll of paper, which is 94 centimeters wide, into equal blocks, with 1 block for each group of students. Use light pencil to mark off each block. Cut off any leftover paper. Each group will show a different type of transportation."

Students split up into groups and then divided the paper into blocks. Mrs. Perez smiled. She had found a way for her students to practice their math and have a good time.

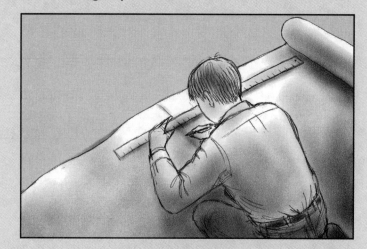

A. How many groups of students were to paint the mural? Use the information from page 60 to complete this multiplication table. Then write your answer below. Write the multiplication and division facts you used to find the solution.

×	1	2	3	4	5	6	7	8	9	10	11	12
divisor												

Write each multiple of the divisor.

Answer: _____

B. How many centimeters wide is each block of the mural? How many centimeters were left over? Use the information from page 60 to complete this flowchart. Then write your answer below.

Divide the tens.	**Bring down the ones digit.**	**Divide the ones.**

Answer: _____

C. Explain your solution to either item A or item B above.

_____

_____

_____

_____

_____

_____

Check Your Understanding

Fill in the letter of the correct answers to questions 1–8.
Write your answers to questions 9 and 10.

1. Stacey picked 39 apples at the orchard. Her mom planned to make 5 apple pies. Stacey divided 39 by 5. What correct quotient did she find?
   - (A) 6 R4
   - (B) 7 R4
   - (C) 8 R1
   - (D) 9 R1

2. Elwood had 82 nickels that he wanted to wrap in 4 rolls and take to the bank. He divided 82 by 4 to find out how many nickels to count for each roll and how many nickels he would have left over. What was his correct answer?
   - (A) 20 R2
   - (B) 21 R1
   - (C) 18 R2
   - (D) 23 R1

3. A Boston radio station gave away 65 tickets to a Red Sox baseball game. Donetta figured out how many pairs of tickets that was. What was her correct answer?
   - (A) 30 R5
   - (B) 33 R2
   - (C) 34 R1
   - (D) 32 R1

4. A rain forest received 58 inches of rain in 6 months. To find about how many inches of rain fell each month, Lois solved 58 ÷ 6. What was her correct answer?
   - (A) 9 R4
   - (B) 14 R2
   - (C) 9 R2
   - (D) 12 R4

5. Lenny's family drove to the beach on Saturday. Because of roadwork and traffic, the 48-mile drive took 2 hours. How many miles did they average each hour?
   - (A) 12 miles
   - (B) 16 miles
   - (C) 14 miles
   - (D) 24 miles

6. Lenny got up early one morning and saw baby sea turtles crawl from their nests in the sand. Lenny counted 5 nests and 90 baby turtles. The same number of baby turtles crawled out of each nest. How many baby turtles crawled from each nest?
   - (A) 14 baby turtles
   - (B) 25 baby turtles
   - (C) 18 baby turtles
   - (D) 15 baby turtles

7. Lenny collected 57 shells at the beach. Back home, he divided the shells equally and arranged them on 3 shelves. How many shells did he put on each shelf?
   - (A) 23 shells
   - (B) 19 shells
   - (C) 15 shells
   - (D) 13 shells

8. Lenny's dad took 7 rolls of film for a total of 84 pictures during the vacation. How many pictures were on each roll of film?
   - (A) 12 pictures
   - (B) 14 pictures
   - (C) 24 pictures
   - (D) 10 pictures

9. After reading about the Empire State Building, Arnie's class built their own model skyscrapers, using craft sticks, toothpicks, and glue. First, the 28 students split up into 7 groups. How many students were in each group?

10. Arnie's teacher asked students to divide 75 packages of craft sticks and 14 boxes of toothpicks evenly among the 7 groups. How many packages of craft sticks and how many boxes of toothpicks did each group get? Were any of the packages of craft sticks or boxes of toothpicks left over? Explain how you found your answer.

Extend Your Learning

- *Divide the Goods*

    Working with a partner, write two division problems for each of these multiplication facts: $2 \times 8 = 16$; $3 \times 5 = 15$; $4 \times 8 = 32$; $5 \times 7 = 35$; $6 \times 2 = 12$; $7 \times 9 = 63$; $8 \times 6 = 48$; $9 \times 4 = 36$; $10 \times 4 = 40$.

- *Art/Social Studies: Create a Mural*

    Plan a mural about a topic you are studying in social studies. Into how many groups of equal numbers of students would your class be split? How big would each section of the mural be? What would be drawn in each section?

STRATEGY SEVEN

Converting Time and Money

Learn About Time

Thinking about the strategy

How do you know how much time has gone by since you started something? How can you figure out how much time you have to get somewhere? You can use a clock face. A clock face can help you find out how much time has passed or how much time remains.

To find out how much time has passed or how much time is left, look at the starting time on the clock face. Then count ahead by hours. Next, count ahead by minutes.

How much time has passed from 3:00 P.M. to 5:30 P.M.?

3:00 P.M.   5:00 P.M.   5:30 P.M.

Studying the problem

Read the problem and the notes beside it.

At what time did they get to the train station?

How many hours and minutes will pass before Stan's grandmother's train arrives?

On a snowy Saturday, Stan and his dad got to the train station at 4:15 P.M. to pick up Stan's grandmother. Stan and his dad soon learned that Stan's grandmother's train was delayed. It was due to arrive in 3 hours, 28 minutes.

At what time will Stan's grandmother's train arrive?

How can Stan use a clock face to solve the problem?

64

Studying the solution

A **clock face** is a graphic organizer that you can use to figure out how much time has passed or what time it will be after a certain amount of time has passed. Stan used these clock faces to figure out the time his grandmother's train would arrive.

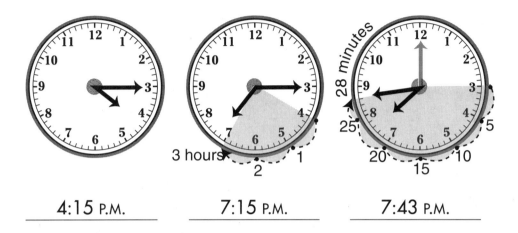

Stan figured out that his grandmother's train would arrive at 7:43 P.M.

Understanding the solution

Read what Stan wrote to explain how he used clock faces to solve the problem.

We arrived at the train station at 4:15 P.M. and learned that my grandmother's train was not due for another 3 hours, 28 minutes. To figure out what time the train would arrive, I drew hour and minute hands to show 4:15 on the first clock face. Next, I counted ahead 3 hours from 4:15 to 7:15. I showed 7:15 on the second clock face. Then I counted minutes. I know that there are 5 minutes between each number on the clock, so I skip-counted by fives from 7:15 to 7:40, which was 25 minutes. Then I counted 3 more minutes from 7:40 to 7:43 to make 28 minutes. I showed 7:43 on the third clock face. I saw that the train would arrive at 7:43 P.M.

Solve a Problem

Studying the problem Read the problem. As you read, think about how you could use clock faces to solve the problem.

Alexa's class is learning about the Pilgrims. The Pilgrims were a group of people who sailed from England to America in 1620 on a ship called the *Mayflower*. After a long, hard journey across the sea, the Pilgrims settled in what is now Plymouth, Massachusetts.

Alexa's teacher, Ms. Adams, told students that on Monday, they were going to Plimoth Plantation in Plymouth. There they would see how people used to live a long time ago. They would also take a tour of a ship built to look just like the real *Mayflower*. Ms. Adams asked students to be at the school on Monday at 8:10 A.M., 40 minutes earlier than they usually arrive at school. The bus would leave at 8:20 A.M. In the past, the ride to Plymouth has taken 1 hour, 35 minutes.

At what time should the bus get to Plymouth?

Finding the solution Use the information from the problem to complete these clock faces. Then write your answer below.

_____   _____   _____

Answer: _____

Explaining the solution

Reread "Understanding the solution" on page 65, which tells how Stan used clock faces to solve a problem. Then write your own explanation of how you completed the clock faces on page 66 and found your solution.

_____

_____

_____

_____

_____

_____

_____

Applying the solution

Use your clock faces on page 66 to answer these questions.

1. What was the last thing you did to find when the students would arrive in Plymouth?

   _____

2. What time did you show on each clock face, from left to right?

   _____

3. At what time will the students arrive in Plymouth if the ride takes 4 minutes longer than expected? _____

4. How much time passed from when the students arrived at school until the time they were to arrive at Plymouth?

   _____

5. If the bus leaves Plimoth Plantation at 1:47 P.M., how much time will the students have spent there? _____

Learn About Money

Thinking about the strategy

To make sure that you get back the right amount of money when you buy something, you need to know how to make change.

A **money table** is a graphic organizer that you can use to find the exact amount of change you should get, with the fewest number of coins and bills.

How much change should Franco receive if he uses a 5-dollar bill to pay for a roll of film that cost $2.56? Franco used this money table to find the answer.

	number	amount
pennies	(4 pennies)	$2.56 + $0.04 = $2.60
nickels	(1 nickel)	$2.60 + $0.05 = $2.65
dimes	(1 dime)	$2.65 + $0.10 = $2.75
quarters	(1 quarter)	$2.75 + $0.25 = $3.00
dollars	(2 dollars)	$3.00 + $2.00 = $5.00

Answer: Franco should receive $0.04 + $0.05 + $0.10 + $0.25 + $2.00, which equals $2.44.

Understanding the solution

Read what Franco wrote to explain how he used a money table to solve the problem.

Count on to find the nearest 5-cent amount.

Count on to find the nearest 10-cent amount.

Count on to find the nearest dollar amount.

Count on to find the nearest 5-dollar amount.

To use the fewest number of coins and bills, I had to use coins and bills with the greatest values, such as quarters and dollars. To find the nearest amount that ended in 0 or 5, I needed 4 pennies, which brought me to $2.60. I saw that I could use 1 nickel to get to $2.65 and 1 dime to get to $2.75. I then skip-counted by twenty-fives to get to $3.00. Next, I counted on from $3.00 to $5.00 and saw that I needed 2 dollars. I added up the values of the coins and bills and got $2.44 as the amount that I should get.

Solve a Problem

Finding the solution

Tess saw a tiny wooden train on a junk table at the spring fair. The man at the junk table told Tess that she could buy the train for $3.37. Tess gave the man a 5-dollar bill. How much change should Tess receive, using the fewest number of coins and bills?

Complete this money table to find how much change Tess should receive. Then write your answer below.

	number	amount
pennies		
nickels		
dimes		
quarters		
dollars		

Answer: _____

_____

Explaining the solution

Write an explanation of how you found the answer by completing the money table above.

_____

_____

_____

_____

_____

Numbers in Context

Read "Mr. Exactly." Think about the ways that numbers are used in the selection. Then answer items A–C on page 71.

Mr. Exactly

Once, there was a man who did everything exactly the same, day after day. His neighbors called him Mr. Exactly.

Mr. Exactly wore the same outfit every day—red plaid pants, a white shirt, a green polka-dotted bow tie, purple socks, and brown shoes. Mr. Exactly's clothes were always clean, and his shoes were always shiny.

Every day, at exactly 7:09 A.M., Mr. Exactly opened his front door, stepped outside, and said, "What shall I do this wonderful, beautiful day?" Anyone who happened to hear him would chuckle. Mr. Exactly did exactly the same thing every single day, rain or shine.

At 7:03 A.M., he ate a bowl of cereal with ice-cold milk. From 7:10 A.M. until 9:24 A.M., he read the paper. At 9:25 A.M., he went for a 28-minute walk in the park. Then he went home, took a 6-minute shower, and got dressed for the day. At 10:05 A.M., he drove to the library, where he worked. His job was to make sure all the books were in exactly the right order.

At 1:11 P.M. every day, Mr. Exactly walked from the library to Suzi's Diner for lunch. He ordered exactly the same lunch every day—a cheese sandwich, an apple, and a glass of milk. His bill was always $3.21. Of course, he always had the exact amount of money to pay the bill.

Then one day, Mr. Exactly reached into his pocket and was shocked. All he had was a 5-dollar bill. "I'll have to bring you change," Suzi said.

Mr. Exactly was not used to change of any kind. As Suzi walked away, he called out, "Of course, you'll bring me the right change."

"Exactly right," Suzi said.

A. How long does Mr. Exactly spend reading the paper each morning? Use the information from page 70 to complete these clock faces. Then write your answer below.

_____   _____   _____

Answer: _____

B. How much change should Mr. Exactly receive from Suzi, using the fewest coins and bills? Use the information from page 70 to complete this money table. Then write your answer below.

	number	amount
pennies		
nickels		
dimes		
quarters		
dollars		

Answer: _____

_____

C. Explain your solution to either item A or item B above.

_____

_____

_____

_____

Check Your Understanding

Fill in the letter of the correct answers to questions 1–8.
Write your answers to questions 9 and 10.

1. One rainy Sunday, Benny and his mom played chess from 1:15 P.M. until 4:20 P.M. How long did they play chess?
   - Ⓐ 2 hours, 20 minutes
   - Ⓑ 3 hours, 35 minutes
   - Ⓒ 5 hours, 15 minutes
   - Ⓓ 3 hours, 5 minutes

2. Yan's third-grade class watched a movie about how oysters make pearls. The movie started at 11:06 A.M. and lasted 47 minutes. At what time did the movie end?
   - Ⓐ 11:43 A.M.
   - Ⓑ 11:53 A.M.
   - Ⓒ 11:41 A.M.
   - Ⓓ 11:50 A.M.

3. Nadia and her dad got on a train to New York City at 2:19 P.M. They arrived 6 hours, 14 minutes later. At what time did they get to New York City?
   - Ⓐ 8:33 P.M.
   - Ⓑ 6:15 P.M.
   - Ⓒ 7:32 P.M.
   - Ⓓ 8:00 P.M.

4. The next day, Nadia and her dad went sightseeing. They began a tour at 12 noon and finished at 6:40 P.M. How long were they on the tour?
   - Ⓐ 5 hours, 40 minutes
   - Ⓑ 5 hours, 20 minutes
   - Ⓒ 6 hours, 40 minutes
   - Ⓓ 6 hours, 20 minutes

5. Mark used a 5-dollar bill to pay for a kite-making kit that cost $3.04. Which coins and bills show the change that he should have received?
   - Ⓐ 1 penny, 3 nickels, 1 dime, 2 quarters, 2 one-dollar bills
   - Ⓑ 1 penny, 2 dimes, 3 quarters, 1 one-dollar bill
   - Ⓒ 1 penny, 3 quarters, 1 one-dollar bill
   - Ⓓ 1 penny, 2 nickels, 2 quarters, 1 one-dollar bill

6. Chee used a 5-dollar bill to pay for yarn and beads that cost $2.74. How much change should she get?
   - Ⓐ $3.36
   - Ⓑ $2.24
   - Ⓒ $2.26
   - Ⓓ $2.00

7. Shirley used a 5-dollar bill to pay for a small bag of peanuts that cost $0.53 and a bottle of water that cost $0.99. How much change should she get?
   - Ⓐ $1.52
   - Ⓑ $4.01
   - Ⓒ $2.46
   - Ⓓ $3.48

8. Bruce used a 1-dollar bill to pay for two raffle tickets that cost $0.33 each. How much change should Bruce get?
   - Ⓐ $0.34
   - Ⓑ $0.67
   - Ⓒ $0.77
   - Ⓓ $1.66

9. Mike and his dad got up at 4:05 A.M. to watch a meteor shower. They were able to see meteors until 6:09 A.M. Then the sun rose, and the sky got too light. How long did Mike and his dad spend watching the meteor shower before the sun came up?

___

10. Luisa used a 5-dollar bill to pay for two subway tokens that cost $1.35 each. What amount of change should Luisa have received? What is the fewest number of coins and bills that she could receive? Explain how you found your answer.

___
___
___
___
___
___
___
___
___

Extend Your Learning

- *School Supply Store*

    Work with a group to set up a classroom store that sells books, pencils, rulers, erasers, and other items. Price everything under $5.00. Using play money, take turns shopping and making change.

- *Reading: Reading Log*

    Keep a weekly reading log. Choose a new book from the library, or reread an old favorite. Each day, jot down the exact time you start reading and the exact time you stop reading. At the end of the week, find how much time you spent reading that week.

STRATEGY EIGHT: Converting Customary and Metric Measures

Learn About Customary Measures

Thinking about the strategy

What is the purpose of these common tools?

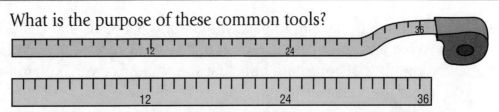

You use these tools to measure lengths in customary measures. Inches, feet, yards, and miles are customary measures of length. You sometimes need to convert, or change, one customary measure of length to another customary measure of length.

Raoul read about the boa constrictor, a kind of snake found in the rain forest. He learned that one snake had grown to 36 feet in length. Raoul used the conversion table below to convert the measurement to inches and then to yards. What operations did he use?

Go from larger units of measure to smaller units.

Multiply	→	By	→	To Get
__36__ feet	×	12	=	__432__ inches
____ yards	×	3	=	____ feet
____ miles	×	5,280	=	____ feet

Go from smaller units of measure to larger units.

Divide	→	By	→	To Get
____ inches	÷	12	=	____ feet
__36__ feet	÷	3	=	__12__ yards
____ feet	÷	5,280	=	____ miles

Studying the problem Read the problem and the notes beside it.

What is the tallest a person can be for the kiddie rides? How tall is Beverly? What should Beverly do to convert inches to feet? Feet to inches?

Beverly read the rules about the kiddie rides at the amusement park. "You must be shorter than 36 inches to ride these rides," one rule said. Beverly was exactly 4 feet tall.

Can Beverly ride any of the kiddie rides?

How can Beverly use a conversion table to solve the problem?

Studying the solution   A **conversion table** is a graphic organizer that you can use to convert customary measures. Beverly used this conversion table to find if she could ride any of the kiddie rides at the amusement park.

Go from larger units of measure to smaller units.

Multiply →	By →	To Get
__4__ feet ×	12 =	__48__ inches
_____ yards ×	3 =	_____ feet
_____ miles ×	5,280 =	_____ feet

Go from smaller units of measure to larger units.

Divide →	By →	To Get
__36__ inches ÷	12 =	__3__ feet
_____ feet ÷	3 =	_____ yards
_____ feet ÷	5,280 =	_____ miles

Beverly was too tall to ride the kiddie rides, because you must be shorter than 36 inches.

Understanding the solution   Read what Beverly wrote to explain how she used a conversion table to solve the problem.

I could solve the problem two ways. I could convert 36 inches to feet, or 4 feet to inches. To go from inches to feet, I have to change from a smaller unit to a larger unit, so I have to divide. I found the row in the table for changing inches to feet. On the line, I wrote 36 inches, the height limit for the kiddie rides. I moved across the row and saw that I had to divide by 12, the number of inches in a foot. I divided by 12 and got 3 feet, which I wrote in the table. I then knew that any child 3 feet or over could not ride the kiddie rides. I am 4 feet tall, too tall to ride the kiddie rides. I also solved the problem by converting my height from feet to inches. To convert from a larger unit to a smaller unit, I have to multiply. I multiplied 4 feet by 12 and got 48 inches. I could see that I am 12 inches (or 1 foot) too tall to ride the kiddie rides.

Solve a Problem

Studying the problem   Read the problem. As you read, think about how you could use a conversion table to solve the problem.

Nate's mom bought Nate a math adventure video game. The game has 12 levels. The goal of the game is to use math skills to find a treasure that has been scattered in 12 different places. Nate has found the first 7 pieces of the treasure. The 8th piece is buried on an island. Nate has to convert customary measures to find it. Here are his directions.

What are his correct conversions?

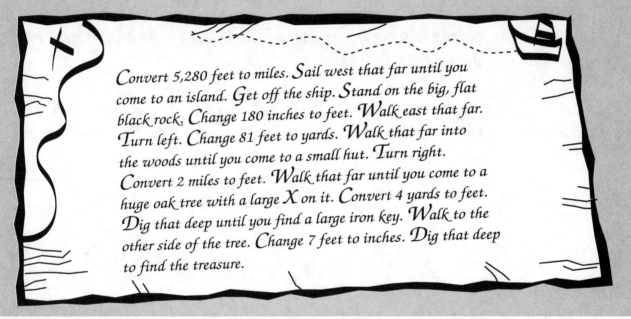

Convert 5,280 feet to miles. Sail west that far until you come to an island. Get off the ship. Stand on the big, flat black rock. Change 180 inches to feet. Walk east that far. Turn left. Change 81 feet to yards. Walk that far into the woods until you come to a small hut. Turn right. Convert 2 miles to feet. Walk that far until you come to a huge oak tree with a large X on it. Convert 4 yards to feet. Dig that deep until you find a large iron key. Walk to the other side of the tree. Change 7 feet to inches. Dig that deep to find the treasure.

Finding the solution   Use the information from the problem to complete this conversion table. Then write your answer below.

Multiply → By → To Get					
_____ feet	×	12	=	_____	inches
_____ yards	×	3	=	_____	feet
_____ miles	×	5,280	=	_____	feet

Divide → By → To Get					
_____ inches	÷	12	=	_____	feet
_____ feet	÷	3	=	_____	yards
_____ feet	÷	5,280	=	_____	miles

Answer: _____

Explaining the solution

Reread "Understanding the solution" on page 75, which tells how Beverly used a conversion table to solve a problem. Then write your own explanation of how you completed the conversion table on page 76 and found your solution.

_____

_____

_____

_____

_____

_____

_____

_____

Applying the solution

Use your conversion table on page 76 to answer these questions.

1. What operation did you use to convert 81 feet to yards?

   _____

2. What operation did you use to convert 4 yards to feet?

   _____

3. In the game, how many miles did the ship sail to reach the island?

   _____

4. The directions say to change 180 inches to feet.

   How many yards is that? _____

5. Would you find the treasure if you dug down 78 inches? Explain your answer.

   _____

   _____

Learn About Metric Measures

Thinking about the strategy

Some things are measured in metric units. Metric measures of length include centimeters (cm), decimeters (dm), meters (m), and kilometers (km). You sometimes need to convert one metric measure of length to another metric measure of length.

A **conversion table** is a graphic organizer that you can use to convert metric measures.

How can Shivonne convert the length of a road 6,000 meters long to decimeters and kilometers? Shivonne used this conversion table to find the answer.

Multiply	→	By	→	To Get
_____ dm	×	10	=	_____ cm
6,000 m	×	10	=	60,000 dm
_____ m	×	100	=	_____ cm
_____ km	×	1,000	=	_____ m

Divide	→	By	→	To Get
_____ cm	÷	10	=	_____ dm
_____ dm	÷	10	=	_____ m
_____ cm	÷	100	=	_____ m
6,000 m	÷	1,000	=	6 km

Answer: Shivonne figured out that the road was 60,000 decimeters, or 6 kilometers, long.

Understanding the solution

Read what Shivonne wrote to explain how she used a conversion table to solve the problem.

Look to see if you are going from a larger unit of measure to a smaller unit, or from a smaller unit of measure to a larger unit.

A decimeter is smaller than a meter.

A kilometer is larger than a meter.

I found the row in the table for changing meters to decimeters. I wrote 6,000 m. I moved across the row and saw that I had to multiply by 10. I multiplied 6,000 m by 10 and got 60,000 dm. Next, I found the row in the table for changing meters to kilometers. I wrote 6,000 m. I moved across the row and saw that I had to divide by 1,000. I divided 6,000 m by 1,000 and got 6 kilometers.

Solve a Problem

Finding the solution

On the Web, Jay read these facts about sea turtles.

1. A full-grown Kemp's ridley sea turtle may measure only 7 dm.
2. Loggerhead turtles average about 1 m in length.
3. Loggerheads nest in Florida on a stretch of beach that is 33 km long.
4. Some loggerheads travel as few as 4,000 m to find food. Others may travel thousands of kilometers.
5. Leatherbacks are the largest sea turtles and may grow to 190 cm.
6. Some leatherbacks dive over 10,000 dm to find jellyfish to eat.

Complete this conversion table. Change the measures in facts 1, 2, and 3 to smaller units. Change the measures in facts 4, 5, and 6 to larger units. Then find about how much longer in dm a leatherback is than a Kemp's ridley. Write your answer below.

Multiply	→	By	→	To Get
_____ dm	×	10	=	_____ cm
_____ m	×	10	=	_____ dm
_____ m	×	100	=	_____ cm
_____ km	×	1,000	=	_____ m

Divide	→	By	→	To Get
_____ cm	÷	10	=	_____ dm
_____ dm	÷	10	=	_____ m
_____ cm	÷	100	=	_____ m
_____ m	÷	1,000	=	_____ km

Answer: _____

Explaining the solution

Write an explanation of how you found the answer by completing the conversion table above.

_____
_____
_____
_____
_____

Numbers in Context

Read "Ronald the Ruler." Think about the ways that numbers are used in the selection. Then answer items A–C on page 81.

Ronald the Ruler

"Hi," the new girl said. "I'm Ann. May I sit with you for lunch?"

"Sure," Pete said, scooting over on the bench. "I'm Pete."

"I'm Jewel, and he's Chan," the girl next to Pete said.

"I'm Ronald," the boy opposite Chan said, "but you can call me 'The Ruler.'"

"Why?" Ann asked. "Do you act like a king?"

"No," Jewel said. "It's because he has this strange need to measure things."

"What kinds of things do you measure?" Ann asked.

"Everything!" Pete, Jewel, and Chan shouted at once.

"Let's see," Ronald said. "I recently measured the length of the new gym. It's 936 inches. Before the gym was built, the school was 100 feet long. Sometimes, I'll measure and convert. For example, I found out the height of the school flagpole in feet and converted that to 20 yards. I measured my height in centimeters and changed that figure to 13 decimeters."

Pete said to Ronald, "Tell Ann what you want for your birthday."

"A laser, digital measuring tape," Ronald said. "It measures in customary or in metric measures. It's so very cool!"

The bell signaling the end of lunch period rang. "I have to get a library card after school," Ann said. "How far is it to the library?"

"It's 8 kilometers from here to the main library," Ronald said, "but there's a small branch library only 3,000 meters from here."

Ann laughed along with her new friends. "I should have known that Ronald the Ruler would know the answer," she said.

A. What is the height of the school flagpole in feet? What is the total length in inches and in feet of the entire school with the new gym building? Use the information from page 80 to complete this conversion table. Then write your answer below.

Multiply → By → To Get	Divide → By → To Get
_____ feet × 12 = _____ inches	_____ inches ÷ 12 = _____ feet
_____ yards × 3 = _____ feet	_____ feet ÷ 3 = _____ yards
_____ miles × 5,280 = _____ feet	_____ feet ÷ 5,280 = _____ miles

Answer: _____

B. How tall in centimeters is Ronald? How much closer in meters and kilometers is it to the branch library than to the main library? Use the information from page 80 to complete this conversion table. Then write your answer below.

Multiply → By → To Get	Divide → By → To Get
_____ dm × 10 = _____ cm	_____ cm ÷ 10 = _____ dm
_____ m × 10 = _____ dm	_____ dm ÷ 10 = _____ m
_____ m × 100 = _____ cm	_____ cm ÷ 100 = _____ m
_____ km × 1,000 = _____ m	_____ m ÷ 1,000 = _____ km

Answer: _____

_____

C. Explain your solution to either item A or item B above.

_____

_____

_____

_____

_____

Check Your Understanding

Fill in the letter of the correct answers to questions 1–8.
Write your answers to questions 9 and 10.

1. Rhoda's bookcase is 36 inches wide. What is the width in feet and in yards?
   - Ⓐ 3 feet, or 1 yard
   - Ⓑ 108 feet, or 324 yards
   - Ⓒ 432 feet, or 5,184 yards
   - Ⓓ 360 feet, or 10 yards

2. Hildie and her dad went on a 4-mile hike. How many feet did they hike?
   - Ⓐ 48 feet
   - Ⓑ 12 feet
   - Ⓒ 21,120 feet
   - Ⓓ 4,000 feet

3. Mike is 60 inches tall. His uncle Jon is 1 foot taller. How tall in inches is Mike's uncle?
   - Ⓐ 721 inches
   - Ⓑ 61 inches
   - Ⓒ 12 inches
   - Ⓓ 72 inches

4. Evan used to live 31,680 feet from the state capital. Now he lives 3 miles from the capital. What is the difference in miles from his new house to his old house?
   - Ⓐ 29 miles
   - Ⓑ 3 miles
   - Ⓒ 877 miles
   - Ⓓ 6 miles

5. The Andean condor, a bird of prey, has a wingspan of more than 3 meters. About how wide in cm and dm is the condor's wingspan?
   - Ⓐ 30 cm, or 300 dm
   - Ⓑ 36 cm, or 9 dm
   - Ⓒ 300 cm, or 30 dm
   - Ⓓ 30,000 cm, or 3,000 dm

6. Josh lives 12 km from the nearest airport. How far is that in meters?
   - Ⓐ 12,000 m
   - Ⓑ 120 m
   - Ⓒ 1,200 m
   - Ⓓ 1.2 m

7. The emperor penguin stands more than 110 cm tall. The fairy penguin stands about 4 dm tall. What is the difference in dm in the two measurements?
   - Ⓐ 106 dm
   - Ⓑ 290 dm
   - Ⓒ 1,096 dm
   - Ⓓ 7 dm

8. Sue and Emma built a snowman. The body was 120 cm tall. The head was 6 dm tall. How tall in cm was the snowman?
   - Ⓐ 18 cm
   - Ⓑ 180 cm
   - Ⓒ 126 cm
   - Ⓓ 720 cm

9. Nine years ago, Louise's grandmother planted a tree in honor of Louise's birth. The tree sapling was only 24 inches tall when it was planted. In 9 years, it has grown 9 feet. How tall in inches and feet is the tree now?

_____

10. Ray's mom wants to put bunk beds in Ray's room. The set-up bunk beds reach a height of 18 dm. The ceiling is 3 m from the floor. At least 100 cm of space must be between the top of the mattress of the top bunk and the ceiling. Is there enough room to fit the bunk beds in Ray's room? Explain how you found your answer.

_____

_____

_____

_____

_____

_____

_____

Extend Your Learning

- *How Far Is It?*

    Find out how far it is from your school to the nearest park, airport, mall, train station, and major highway. Convert each number to a smaller or larger unit of measure.

- *Science: Compare Animal Traits*

    Work with a group to research two kinds of the same animal. For example, a lion and a tiger; a rattlesnake and a cobra; or an eagle and a hawk. Look for number facts that answer questions about each animal such as: How long is it? How tall is it? How fast can it move? How far does it travel to nest or to find food? Practice converting these number facts into smaller or larger units of measure. Then compare measures.

STRATEGY NINE: Using Algebra

Learn About Algebra

Thinking about the strategy

Numbers can be used to make patterns. Patterns follow rules. The rules help you to continue the patterns.

A function machine can help you continue a pattern. A number goes into the machine. A function is done to the number, and a new number comes out. The function is the rule.

There are 5 baseball cards in a package. How many cards will you get if you buy different numbers of packages? This function machine gives you the answer. The rule is "multiply by 5," because there are 5 cards in a package. A number goes into the machine. The rule is used on the number. A new number comes out.

Rule: Multiply by 5.	
In	Out
1	5
2	10
3	15
4	20

(1 x 5 = 5)
(2 x 5 = 10)
(3 x 5 = 15)
(4 x 5 = 20)

Studying the problem Read the problem and the notes beside it.

How old is Erica?

What rule can you use to find Daniel's age?

Daniel is 3 years older than his sister Erica. Erica is 9 years old. How old is Daniel? How old will Daniel and Erica each be over the next 5 years?

How can Erica use a function machine to solve the problem?

Studying the solution A **function machine** is a graphic organizer that you can use to continue a pattern. Erica used this function machine to find hers and Daniel's ages. The rule she used is "add 3."

Rule: Add 3.	
In (Erica's Age)	Out (Daniel's Age)
9	12
10	13
11	14
12	15
13	16
14	17

Daniel is now 12 years old. The function machine shows Erica's and Daniel's ages now and over the next 5 years.

Understanding the solution Read what Erica wrote to explain how she used a function machine to solve the problem.

Since Daniel is 3 years older than I, the rule is to add 3 to my age. I set up a function machine. I wrote the rule "add 3" at the top. The In column is my age (Erica's age). The Out column is Daniel's age. Since I am 9 now, I put 9 in the first row. I added 3 to 9 to find Daniel's age (12). Next, I wrote my age for the next 5 years (10, 11, 12, 13, 14). I used the rule to find Daniel's age for each of these years.

$$10 + 3 = 13$$
$$11 + 3 = 14$$
$$12 + 3 = 15$$
$$13 + 3 = 16$$
$$14 + 3 = 17$$

Solve a Problem

Studying the problem   Read the problem. As you read, think about how you could use a function machine to solve the problem.

Pablo likes to use his computer. His printer can print about 4 pages in 1 minute. Pablo wants to know many pages the printer will print in different numbers of minutes.

What rule can Pablo use to find the answer?

Finding the solution   Use the information from the problem to complete this function machine. Then write your answer below.

Rule: _____	
In (Number of Minutes)	**Out** (Number of Pages)
1	
2	
3	
4	
5	
6	

Answer: _____

Explaining the solution

Reread "Understanding the solution" on page 85, which tells how Erica used a function machine to solve a problem. Then write your own explanation of how you completed the function machine on page 86 and found your solution.

_____

_____

_____

_____

_____

_____

Applying the solution

Use your function machine on page 86 to answer these questions.

1. What is the pattern of the numbers in the In column?

   _____

   _____

2. What is the pattern of the numbers in the Out column?

   _____

   _____

3. How long will it take the printer to print 20 pages?

   _____

4. How many pages can be printed in 3 minutes?

   _____

5. How can you use the rule to find how many pages can be printed in 9 minutes?

   _____

Learn More About Algebra

Thinking about the strategy

A **grid** is a graphic organizer that you can use to show where items are located.

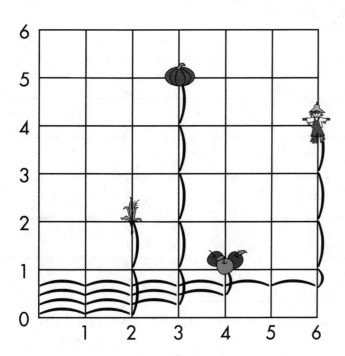

Every year in the fall, the city entertainment committee hires Mr. Marino. He makes a huge maze out of hay bales. The maze has some fall items in different places. Mr. Marino used this grid to show where the items are. How can Mr. Marino use coordinate pairs to name the locations of the items?

Answer: Mr. Marino found the coordinate pairs to be: cornstalk (2, 2); pumpkin (3, 5); apples (4, 1); scarecrow (6, 4).

Understanding the solution

Read what Mr. Marino wrote to explain how he used a grid to solve the problem.

Count the spaces starting from the left side.

Count the spaces starting from the bottom at 0.

The first number shows how many spaces over.

The second number shows how many spaces up.

    The grid has numbers that label each line going from left to right. There are also numbers for each line that goes up and down. There are many points where these lines meet. The points can be named with two numbers, called a coordinate pair.

I started with the cornstalk. From the 0, I moved over 2 spaces and then up 2 spaces. This point has the coordinate pair (2, 2). The first number shows how many spaces you move over from 0. The second number shows how many spaces you move up from 0. The apples are over 4 and up 1: (4, 1). The scarecrow is over 6 and up 4: (6, 4). The pumpkin is over 3 and up 5: (3, 5).

Solve a Problem

Finding the solution

Charlotte works at a craft store. She is making a display of small wooden objects. She has already placed the flower, shamrock, and star. She has directions that tell where the moon, triangle, and heart are to be put. Where should Charlotte draw the moon, triangle, and heart? What are the coordinate pairs of the flower, shamrock, and star?

Complete this grid by drawing the moon, triangle, and heart where they belong. Find the coordinate pairs for the flower, shamrock, and star. Then write your answer below.

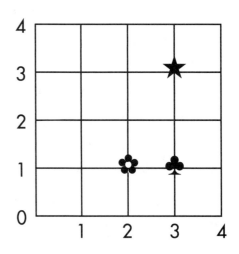

Answer: _____

Explaining the solution

Write an explanation of how you found the answer by completing the grid above.

_____

_____

_____

_____

_____

_____

Numbers in Context

Read "The Big Cleanup." Think about the ways that numbers are used in the selection. Then answer items A–C on page 91.

The Big Cleanup

In Frost Valley, the winters are very hard and long. The people who live in Frost Valley look forward to spring. That's why everyone helps out at the town's spring cleanup.

The cleanup planning group makes a map of the downtown area. They use a grid to show the main work areas. The park is at (1, 5) on the grid. The town square is at (3, 3). The elementary school is at (5, 2). The middle school and high school are at (6, 5).

All of the cleanup work takes supplies and money. Many people visit the businesses in town. Some businesses give the supplies that are needed. Others give money. Money is raised in other ways too. A car wash is held at the Frost Valley High School. The charge for a car wash is $3.00. Everyone hopes that many people will show up to have their cars washed.

One group of workers is in charge of planting flowers. The Frost Valley Nursery donates all of the flowers. The workers plant the flowers at different places around town. Each person in the group gets 8 flowers to plant.

At the end of the cleanup, Frost Valley looks beautiful once again.

A. How many flowers can be planted if there are 5 workers? How much money can be raised by washing 4 cars? Use the information from page 90 to complete these function machines. Then write your answer below.

Rule: _____	
In (Number of Workers)	**Out** (Number of Flowers)
1	
2	
3	
4	
5	
6	

Rule: _____	
In (Number of Cars)	**Out** (Amount of Money)
1	
2	
3	
4	
5	
6	

Answer: _____

B. Use the letters to label the following items on the grid: *A*—park; *B*—town square; *C*—elementary school; *D*—middle school/high school. How would you move to get from the town square to the high school? Use the information from page 90 to plot the points on this grid. Then write your answer below.

Answer: _____

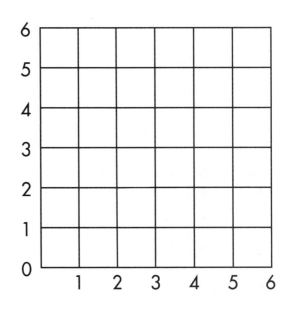

C. Explain your solution to either item A or item B above.

_____

_____

_____

_____

Check Your Understanding

Fill in the letter of the correct answers to questions 1–8.
Write your answers to questions 9 and 10.

Use this grid to answer questions 1–4.

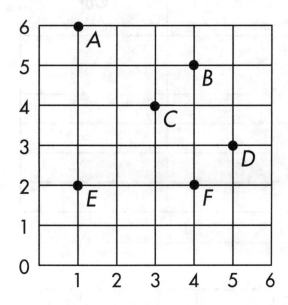

1. What point is named by the coordinate pair (4, 2)?
   - Ⓐ F
   - Ⓑ E
   - Ⓒ D
   - Ⓓ C

2. What coordinate pair names point *B*?
   - Ⓐ (5, 6)
   - Ⓑ (5, 5)
   - Ⓒ (5, 4)
   - Ⓓ (4, 5)

3. What point is named by the coordinate pair (3, 4)?
   - Ⓐ A
   - Ⓑ B
   - Ⓒ C
   - Ⓓ D

4. What coordinate pair names point *E*?
   - Ⓐ (4, 2)
   - Ⓑ (2, 1)
   - Ⓒ (1, 2)
   - Ⓓ (2, 2)

Use this function machine to answer questions 5 and 6.

Rule: Add 6.	
In	Out
1	7
2	8
?	?

5. If the number 4 goes into the In column, what number goes into the Out column?
   - Ⓐ 9
   - Ⓑ 10
   - Ⓒ 11
   - Ⓓ 12

6. The number 21 is in the Out column. What number is in the In column?
   - Ⓐ 15
   - Ⓑ 16
   - Ⓒ 26
   - Ⓓ 27

7. Brit made cookies for a bake sale. She put 3 cookies into each bag. If she made 8 bags, how many cookies are there in all?
   - Ⓐ 11 cookies
   - Ⓑ 18 cookies
   - Ⓒ 21 cookies
   - Ⓓ 24 cookies

8. Each cup holds 8 ounces of lemonade. How much lemonade is needed to fill 6 cups?
   - Ⓐ 56 ounces
   - Ⓑ 48 ounces
   - Ⓒ 40 ounces
   - Ⓓ 14 ounces

9. At a garage sale, all paperback books have a price of 5¢. What is the cost to buy 5 books? 6 books? 7 books? 8 books? 9 books? 10 books?

10. A game board is set up as a grid. A player started at (0, 0) and went over 2 spaces and up 4 spaces. On her next turn, she moved over 3 spaces and up 3 spaces. What is the coordinate pair of the place she landed on the first move? The second move? Explain how you found your answer.

Extend Your Learning

- *Grid Designs*

    Draw a geometric shape on a grid. Use coordinate pairs to label the corners of the shape. Without showing your partner, name the coordinate pairs, and ask him or her to plot and connect the points. Compare your drawing with your partner's.

- *Physical Education: Jumping-Jack Sets*

    In physical education class, ask your teacher to have a group of students do jumping jacks in sets of 3, and to have another group do jumping jacks in sets of 4. Find how many total jumping jacks both groups do in 1, 2, 3, 4, 5, and 6 sets of exercises.

STRATEGY TEN: Using Geometry

Learn About Geometry

Thinking about the strategy

By using geometry, you can measure how much space a figure takes up. When you do this, you are finding the area.

A rug measures 6 feet by 9 feet. How much area does the rug take up?

You can use a grid to find the area. The grid is made up of squares that are all the same size. In this grid, each square stands for 1 foot by 1 foot. A square that is 1 foot by 1 foot is 1 square foot. The rug is 6 feet wide and 9 feet long. The grid needs to be 6 squares wide and 9 squares long. How can you use the grid to find the area of the rug?

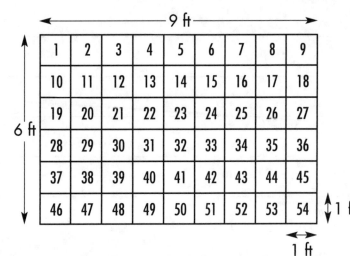

You can count the squares to find the area. You can also add. There are 6 rows with 9 squares in each row: 9 + 9 + 9 + 9 + 9 + 9 = 54. There are 9 columns with 6 squares in each column: 6 + 6 + 6 + 6 + 6 + 6 + 6 + 6 + 6 = 54. The area is 54 square feet.

Studying the problem Read the problem and the notes beside it.

What unit is used to measure the plank?

What is the length?

What is the width?

Martinez works for a roofing company. He and the other workers use many kinds of planks. Some planks are long, and others are shorter. One plank that Martinez uses is 8 feet long and 2 feet wide. Another plank is 16 feet long and 1 foot wide.

What is the area of each plank?

How can Martinez use a grid to solve the problem?

94

Studying the solution A **grid** is a graphic organizer that you can use to find area. Martinez used this grid to find the area of the planks.

		1	2	3	4	5	6	7	8						
		9	10	11	12	13	14	15	16						
1	2	3	4	5	6	7	8	9	10	11	12	13	14	15	16

Martinez found that the area of each plank is 16 square feet.

Understanding the solution Read what Martinez wrote to explain how he used the grid to solve the problem.

The measurements of the planks are given in feet, so I wanted the squares in the grid to stand for feet. I made each square on the grid stand for 1 foot by 1 foot, or 1 square foot. One plank is 8 feet long and 2 feet wide. I drew a rectangle 8 squares long and 2 squares wide. To find the area, I counted the squares. There are 16 squares inside the rectangle, and each square stands for 1 square foot. The area is 16 square feet.

The other plank is 16 feet long and 1 foot wide. I drew a rectangle 16 squares long and 1 square wide. I then counted the squares. There are 16 squares inside the rectangle, so the area is 16 square feet.

Solve a Problem

Studying the problem Read the problem. As you read, think about how you could use a grid to solve the problem.

Laura is helping to build a new house. She is putting plywood down for the floors. A whole sheet of plywood is 4 feet wide and 8 feet long. Laura wants to find how many sheets of plywood are needed. She decided to find the area that 1 sheet will cover.

What is the area of 1 sheet of plywood?

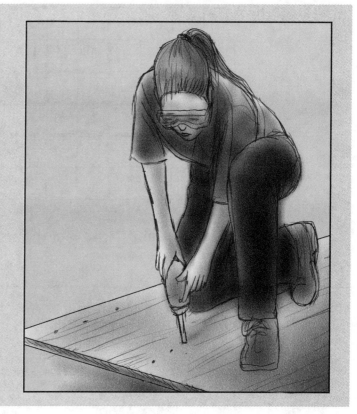

Finding the solution Use the information from the problem to complete this grid. Then write your answer below.

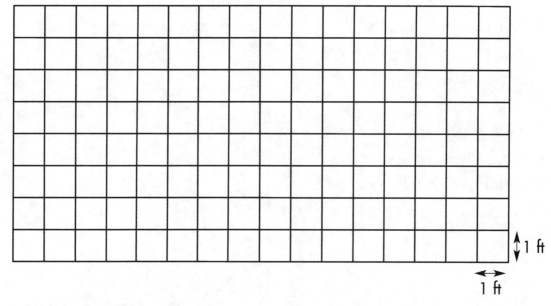

Answer: _____

Explaining the solution

Reread "Understanding the solution" on page 95, which tells how Martinez used a grid to solve a problem. Then write your own explanation of how you completed the grid on page 96 and found your solution.

_____

_____

_____

_____

_____

_____

_____

Applying the solution

Use your grid on page 96 to answer these questions.

1. How many rows of squares are in your rectangle?

   _____

2. How many squares are in each row?

   _____

3. How can you use addition to find the area?

   _____

4. Suppose you place 2 sheets of plywood side by side. How much area would they cover?

   _____

5. How much area would 4 sheets of plywood cover?

   _____

Learn More About Geometry

Thinking about the strategy

You may sometimes want to know how much space a solid figure takes up. This space is called volume. You can divide a solid figure up into cubes that are the same size. Then you can count to see how many cubes fit inside.

A **cube** is a graphic organizer that you can use to find the volume of a solid figure.

Bryan's younger sister has a box filled with blocks. The blocks are all cubes, and they are all the same size. There are 3 layers of blocks in the box. Each layer has 4 rows of 4 blocks. How can Bryan find the volume of the box? Bryan used these cubes to find the answer.

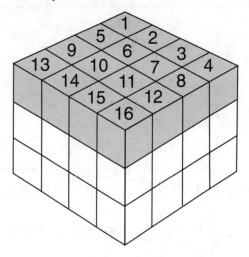

Answer: Bryan can count the blocks to find the volume. There are 48 blocks. The volume of the box is 48 cubic units.

Understanding the solution

Read what Bryan wrote to explain how he used cubes to solve the problem.

Count the cubes in each layer.

Add the number of cubes in each layer.

Give the volume in cubic units.

I can see the cubes in the top layer, so I can count how many. There are 16 blocks in the top layer. Each layer has the same number of cubes. I can add to find out how many are in all three layers: 16 + 16 + 16 = 48. There are 48 cubes in the box. The volume of a box is given in cubic units. So, the volume of the box is 48 cubic units.

Solve a Problem

Finding the solution

Bryan's sister used the blocks to make different figures. The diagram below shows one of her designs. What is the volume of this figure?

Use these cubes to find the volume of the figure. Then write your answer below.

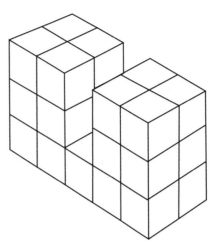

Answer: _____

Explaining the solution

Write an explanation of how you found the answer by using the cubes above.

_____

_____

_____

_____

_____

_____

_____

_____

Numbers in Context

Read "Math Class Cubes." Think about the ways that numbers are used in the selection. Then answer items A–C on page 101.

Math Class Cubes

Mr. Carney's math class is learning about area and volume. They are using connecting cubes to make different figures. Mr. Carney explained that a cube is a solid figure. He told the class that there are 6 faces on a cube, and that all faces are squares of the same size.

The students practiced putting together the cubes to form different shapes. The cubes can connect to others on all 6 sides. Mr. Carney showed how to put cubes together to form one layer. He used 6 rows of cubes to make a bottom layer. There are 6 cubes in each of these rows.

Mr. Carney next connected a second layer of cubes. This layer includes 4 rows of cubes. Each row has 4 cubes in it. To make the top layer, Mr. Carney used 2 rows with 2 cubes in each row.

He displayed the cube tower in front of the class. Then Mr. Carney asked students to find the area of the base. He told them to imagine the square faces of the cubes on the bottom layer of the figure. These squares are used to find the area of the base.

Mr. Carney also asked the class to find the volume of the figure. He reminded the class that area is given in square units and volume is given in cubic units. Then Mr. Carney invited the students to make figures of their own. Students found the area and volume of their own figures. Mr. Carney gave out prizes to students who were correct.

A. What is the area of the first layer, or base, of the figure that Mr. Carney made? Use the information from page 100 to complete this grid. Then write your answer below.

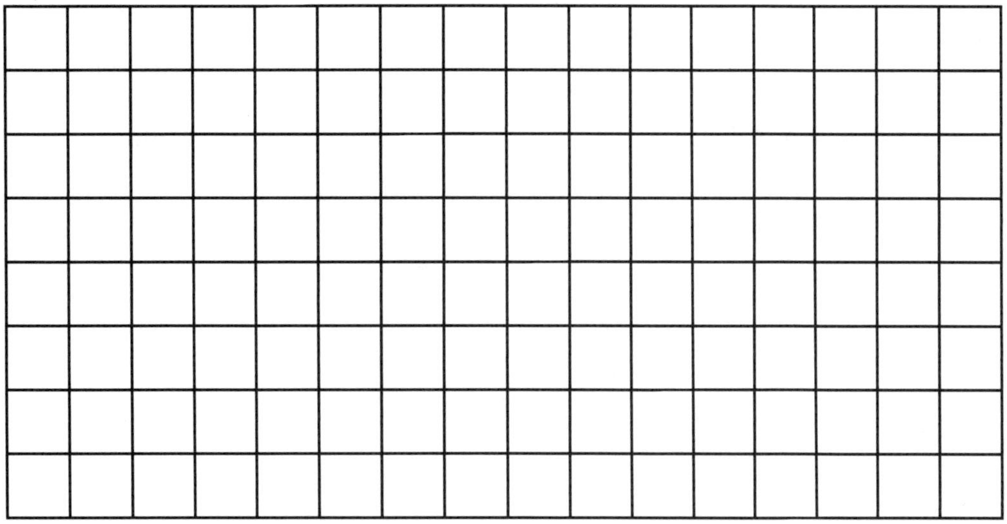

Answer: _____

B. What does Mr. Carney's figure look like? What is the volume of Mr. Carney's figure? Use the information from page 100 to find which figure below is Mr. Carney's. Circle the figure. Then write your volume below.

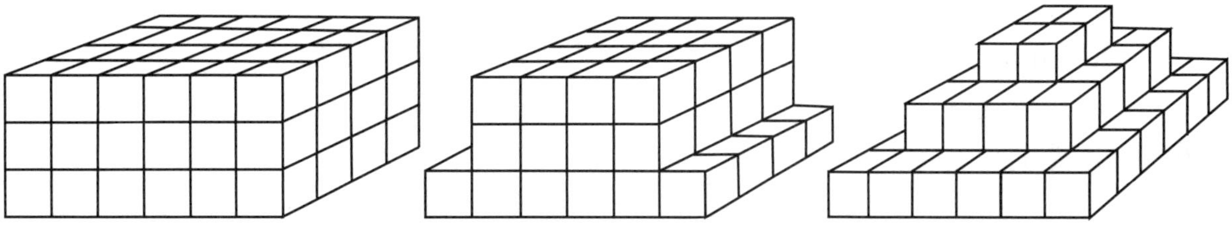

Answer: _____

C. Explain your solution to either item A or item B above.

_____

_____

_____

_____

Check Your Understanding

Fill in the letter of the correct answers to questions 1–8.
Write your answers to questions 9 and 10.

Use this figure to answer questions 1 and 2.

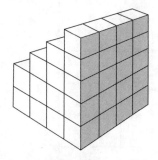

1. What is the volume of the figure?
   - Ⓐ 80 cubic units
   - Ⓑ 60 cubic units
   - Ⓒ 56 cubic units
   - Ⓓ 52 cubic units

2. What is the area of the shaded face in the figure?
   - Ⓐ 24 square units
   - Ⓑ 20 square units
   - Ⓒ 16 square units
   - Ⓓ 14 square units

3. A rectangle has sides that measure 3 inches and 5 inches. What is the area of the rectangle?
   - Ⓐ 2 square inches
   - Ⓑ 8 square inches
   - Ⓒ 13 square inches
   - Ⓓ 15 square inches

4. The area of a rectangle is 72 square feet. The width and length of its sides could be
   - Ⓐ 6 feet by 9 feet.
   - Ⓑ 2 feet by 7 feet.
   - Ⓒ 8 feet by 9 feet.
   - Ⓓ 12 feet by 3 feet.

5. What is the area of this figure?

   - Ⓐ 25 square units
   - Ⓑ 30 square units
   - Ⓒ 36 square units
   - Ⓓ 20 square units

6. What is the volume of a tower with 4 cubes in each layer and 6 layers?
   - Ⓐ 25 square units
   - Ⓑ 16 cubic units
   - Ⓒ 10 cubic units
   - Ⓓ 24 cubic units

Use this figure to answer questions 7 and 8.

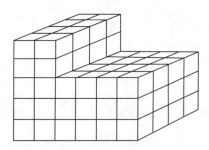

7. What is the area of the base of the figure?
   - Ⓐ 20 square units
   - Ⓑ 24 square units
   - Ⓒ 36 square units
   - Ⓓ 40 square units

8. What is the volume of the figure?
   - Ⓐ 88 square units
   - Ⓑ 60 cubic units
   - Ⓒ 68 cubic units
   - Ⓓ 88 cubic units

9. Two patios are made from blocks that are 1 square foot each. Patio A is 10 feet by 12 feet. Patio B is 12 feet by 16 feet. What patio has the greater area?

_____
_____
_____

10. Look at this figure. Find the volume. Describe the pattern of the blocks in each layer.

_____
_____
_____
_____
_____

Extend Your Learning

- *Connecting Cubes*

    Use connecting cubes to make many different solid figures. Count the cubes in each figure, and record the volume.

- *Science: A Lion's Territory*

    Many animals live inside of their own territory. A lion's territory can be as large as 150 square miles! Suppose that a lion has a territory of 100 square miles. Make a drawing on a grid to show what this territory could look like. (Hint: Have each square on the grid stand for 1 square mile.)

STRATEGY ELEVEN: Determining Probability and Averages

Learn About Probability

Thinking about the strategy

Sometimes, you estimate the chance that something will happen.

Reese looked at this spinner. He predicted that it is more likely that a person will spin the color red than any other color. Red has 2 equal sections on the spinner. The other colors each have only 1 section. Reese decided to do an experiment to test this prediction. He spun the spinner 50 times and recorded the results in a tally chart.

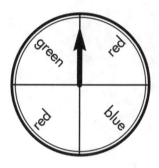

Color	Number of Times Spun
red	|||| |||| |||| |||| |||
green	|||| |||| ||||
blue	|||| |||| |||

In the experiment, what color was spun more than any other color?

Studying the problem

Read the problem and the notes beside it.

How many different colors are there?

How many times was blue picked?

How many times were other colors picked?

    Ms. Lui's class has a box with one each of the following color markers: red, orange, black, green, blue, pink, purple, yellow. If a student picks a marker from the box without looking, Ms. Lui predicts that there is an equally likely chance of picking any color. She decided to test the prediction. Students took turns picking a marker and calling out the color, as shown below. After picking a marker, each student returned the marker to the box.
    Was each color picked about the same number of times?

pink	black	green	yellow	purple	purple	yellow	orange
blue	red	red	orange	black	green	blue	pink
yellow	purple	black	orange	red	green	green	pink
pink	blue	purple	yellow	red	black	blue	pink
purple	orange	green	purple	yellow	yellow	green	black
orange	pink	blue	green	yellow	red	green	orange
purple	yellow						

How can Ms. Lui use a tally chart to solve the problem?

104

Studying the solution   A **tally chart** is a graphic organizer that you can use to record the results of an experiment. Ms. Lui used this tally chart to record the number of times each color was picked. This is one way to check the prediction that it is equally likely that each color is picked.

Color	Number of Times	Number
red	ⅲⅠⅠ	5
orange	ⅲⅠⅠ Ⅰ	6
black	ⅲⅠⅠ	5
green	ⅲⅠⅠ ⅠⅠⅠ	8
blue	ⅲⅠⅠ	5
pink	ⅲⅠⅠ Ⅰ	6
purple	ⅲⅠⅠ ⅠⅠ	7
yellow	ⅲⅠⅠ ⅠⅠⅠ	8

The experiment showed that each color was picked about the same number of times.

Understanding the solution   Read what Ms. Lui wrote to explain how she used a tally chart to solve the problem.

I made a chart and listed all of the different color markers in the first column. In the second column, I recorded the number of times the color was picked. Each time that a student picked a marker, I put a tally mark in the column for the color of the marker. I continued to record the colors in this way. I counted each group of tally marks and then wrote a number in the third column. I had the students pick a marker 50 times. I found the sum of all the numbers to make sure the total was 50.

I looked at the numbers for each color. The least number is 5, and the greatest is 8. There is not a big difference between these numbers. The numbers are all either a 5, 6, 7, or 8. The prediction that there is an equally likely chance of picking any color makes sense with the results of this experiment.

Solve a Problem

Studying the problem Read the problem. As you read, think about how you could use a tally chart to solve the problem.

The following number cards were put into a bag.

| 2 | 3 | 1 | 2 | 2 | 0 | 3 | 4 | 2 | 1 |

What number (or numbers) is most likely to be picked? What number (or numbers) is least likely to be picked? Make a prediction, and then do an experiment. Compare the results of the experiment with your prediction. Make the number cards, and put them into a bag. Pick a card, record the number, and then put the card back into the bag. Repeat this 50 times.

Finding the solution Use the information from the problem to complete this tally chart. Then write your answer below.

Number	Number of Times
0	
1	
2	
3	
4	

Answer: _____

Explaining the solution

Reread "Understanding the solution" on page 105, which tells how Ms. Lui used a tally chart to solve a problem. Then write your own explanation of how you made a prediction and then completed the tally chart on page 106 and found your solution.

_____

_____

_____

_____

_____

_____

_____

Applying the solution

Use your tally chart on page 106 to answer these questions.

1. What would you predict about the chance of picking a 0 or a 4?

   _____

2. What would you predict about the chance of picking a 1 or a 3?

   _____

3. Are you more likely to pick a 0 or a 1?

   _____

4. Are you less likely to pick a 4 or a 3?

   _____

5. Look at your completed tally chart. Write the card numbers in order. Start with the number that was picked the least. End with the number that was picked the most.

   _____

Learn About Averages

Thinking about the strategy

Miki practices her multiplication facts every day. On Monday, she practiced for 5 minutes. On Tuesday, she practiced for 8 minutes. On Wednesday, Miki worked on her facts for 6 minutes. She practiced for 5 minutes on Thursday and for 6 minutes on Friday.

This problem shows that data can change over time. But, you can find an average for the set of data. The average is a number somewhere in the middle of the data set.

A **flowchart** is a graphic organizer that you can use to find the average.

How can Miki find the average amount of time she spent practicing multiplication facts each day? Miki used this flowchart to find the answer.

| $5 + 8 + 6 + 5 + 6 = 30$ | ➡ | $30 \div 5 = 6$ |

1. Find the total number of items in all groups. (Add all the numbers.)

2. Put the total number of items into equal groups. Make the same number of groups that you started with. (Divide the total by the number of groups.)

Answer: Miki practiced her multiplication facts for an average of 6 minutes each day.

Understanding the solution

Read what Miki wrote to explain how she used a flowchart to solve the problem.

Add the numbers in the data set.

Count to find how many pieces are in the data set.

Divide the sum by the number of pieces.

In this data, there are 5 groups. A group is a day of the week. I started by adding the numbers given for each day. The sum is 30. Since there are 5 groups, I have to put the total of 30 into 5 equal groups. To do this, I divided 30 by 5. Since I know that $5 \times 6 = 30$, then $30 \div 5 = 6$. The average time that I spent on multiplication facts each day was 6 minutes.

Solve a Problem

Finding the solution

Alicia plays baseball in the farm league. There are six teams that play at this level. This chart shows the number of girls who are on each team. What is the average number of girls on a team in this league?

Team	Twins	Orioles	Cardinals	Yankees	White Sox	Reds
Number of Girls	2	4	3	3	4	2

Complete this flowchart to find the average number of girls on a team. Then write your answer below.

[ ] ▶ [ ]

1. Find the total number of items in all groups. (Add all the numbers.)

2. Put the total number of items into equal groups. Make the same number of groups that you started with. (Divide the total by the number of groups.)

Answer: _____

Explaining the solution

Write an explanation of how you found the answer by completing the flowchart above.

_____
_____
_____
_____
_____
_____
_____

Numbers in Context

Read "The Spring Fair." Think about the ways that numbers are used in the selection. Then answer items A–C on page 111.

The Spring Fair

Everyone in Springdale looks forward to the spring fair. For as long as anyone can remember, this fair has taken place at the end of April. People say that the spring fair has been going on for so long that the town took its name from the event!

The spring fair is like a kick-off to the warm weather. People can't wait to get outdoors. This year should be the best fair ever. The day will start out with the baseball parade. After the parade, everyone will head to the field. There will be games, food, music, and fun all day long. At dusk, the fireworks will begin.

At the ticket booth, people will be able to buy different kinds of tickets. A ticket for $3.00 will let you play 10 games. With a $4.00 ticket, you can play all the games you want. A $5.00 ticket is good for all games and one chance to win the door prize. If you buy a ticket for $8.00, you can take part in all games and events at the fair.

There is a prize wall at the fair. Children under 10 get to pick a prize from the wall, but they will be blindfolded so that they cannot see the prize they are picking. This just adds to the fun!

A. A child is blindfolded in front of the prize wall. What prize is the child most likely to pick? Write your prediction below. Then do an experiment. Write the names of the prizes on slips of paper. Use the information from page 110 to see how many of each prize to use. Pick a slip of paper, record the prize, and put the paper back into the bag. Do this 50 times, and complete the tally chart. Do the results of your experiment agree with your prediction?

Prize	Number of Times
pencil	
free drink	
book	
grab bag	

Answer: _____

_____

B. What is the average ticket price at the fair? Use the information from page 110 to complete this flowchart. Then write your answer below.

	➤	

1. Find the total number of items in all groups. (Add all the numbers.)

2. Put the total number of items into equal groups. Make the same number of groups that you started with. (Divide the total by the number of groups.)

Answer: _____

C. Explain your solution to either item A or item B above.

_____

_____

_____

_____

_____

Check Your Understanding

Fill in the letter of the correct answers to questions 1–8.
Write your answers to questions 9 and 10.

Use this spinner to answer questions 1 and 2.

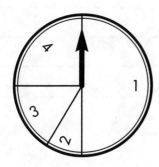

1. What number are you most likely to spin on the spinner?
   - Ⓐ 1
   - Ⓑ 2
   - Ⓒ 3
   - Ⓓ 4

2. What number are you least likely to spin on the spinner?
   - Ⓐ 1
   - Ⓑ 2
   - Ⓒ 3
   - Ⓓ 4

3. What is the average of these numbers?

3	5	4	3	5	4

   - Ⓐ 24
   - Ⓑ 5
   - Ⓒ 4
   - Ⓓ 2

4. Taylor is a scout. The numbers below show how many badges each girl in her troop earned this year. What is the average number of badges earned by each girl this year?

2	3	5	4	2	2

   - Ⓐ 2 badges
   - Ⓑ 3 badges
   - Ⓒ 4 badges
   - Ⓓ 18 badges

Use these number cards to answer questions 5 and 6.

1	2	3	4	5

5. If you pick a number card from the group without looking, which of the following are you most likely to pick?
   - Ⓐ a number greater than 1
   - Ⓑ the number 5
   - Ⓒ the number 3
   - Ⓓ a number less than 3

6. If you pick a number card from the group without looking, which of the following are you least likely to pick?
   - Ⓐ a number less than 4
   - Ⓑ a number greater than 2
   - Ⓒ a number less than 3
   - Ⓓ a number greater than 4

7. The amounts below are prices for souvenirs at a basketball game. What is the average price of a souvenir?

$5.00	$6.00	$7.00	$10.00

   - Ⓐ $28.00
   - Ⓑ $8.00
   - Ⓒ $7.00
   - Ⓓ $6.00

8. The numbers below show the points that Drew scored in the last five basketball games. What is the average number of points that Drew scored in a game?

6	9	8	10	7

   - Ⓐ 9 points
   - Ⓑ 8 points
   - Ⓒ 7 points
   - Ⓓ 6 points

9. Use three different numbers to complete this spinner. Complete the spinner so that you are equally likely to spin all three numbers.

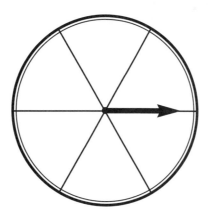

10. In gym class, the green team scored the most points. The players scored the following points: 1, 3, 2, 1, 2, 3. What is the green team's point average? Explain how you found your answer.

_____

_____

_____

_____

_____

Extend Your Learning

- *Rolling Cubes*

    Suppose that you roll two number cubes and add the numbers. Do you know what sum is the most likely to be rolled? It is 6, 7, or 8, because there are three ways to make each of these sums. Conduct an experiment to see what sum is rolled most often. Roll the number cubes 50 times. Record the sums in a tally chart. The possible sums are 2, 3, 4, 5, 6, 7, 8, 9, 10, 11, and 12.

- *Language Arts: Lengths of Sentences*

    How long, do you think, is an average sentence? Use your reading book. Select any 10 sentences from the book. Count and record the number of words in each sentence. Find the average number of words in a sentence.

STRATEGY TWELVE: Interpreting Graphs and Charts

Learn About Graphs

Thinking about the strategy

A line graph can be used to show changes over time. You can show changes from day to day, from week to week, from month to month, and so on.

You may sometimes want to show how something grows. How can you use this line graph to tell how much a tree grew from year 1 to year 4?

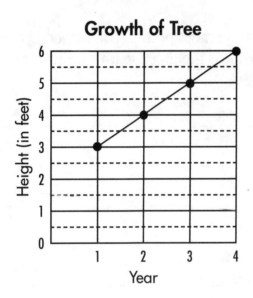

The line graph shows how tall the tree was each year for four years. In year 1, the tree was 3 feet tall. In year 4, it was 6 feet tall. If you subtract 3 from 6 (6 – 3), you will find that the tree grew 3 feet from year 1 to year 4.

Studying the problem Read the problem and the notes beside it.

What is the least number?

What is the greatest number?

What can you use for a scale?

Each week, one student in room 126 is in charge of the lunch count. This week is Nita's turn. On Monday, 8 students bought lunch. The number of students who bought lunch on Tuesday was 10. On Wednesday, 9 students bought lunch, and 12 students bought lunch on Thursday. On Friday, 15 students bought lunch.

On what day did the greatest number of students buy lunch?

How can Nita use a line graph to solve the problem?

Studying the solution   A **line graph** is a graphic organizer that you can use to show changes in data. Nita used this line graph to show how many students bought lunch during one week.

Lunches Bought in Room 126

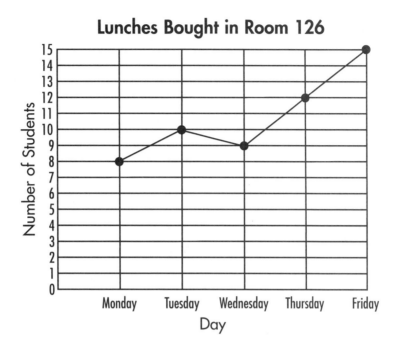

Nita found that the greatest number of students bought lunch on Friday.

Understanding the solution   Read what Nita wrote to explain how she used a line graph to solve the problem.

> To decide on a scale, I looked at the numbers in the data set. The numbers go from 8 to 15. I decided to use a scale with an interval of 1. I started at 0 and drew lines across, up to 15. I labeled the lines 0 to 15. Next, I made a vertical line for each day of the week. I labeled the lines Monday, Tuesday, Wednesday, Thursday, and Friday. Then I labeled the axes and wrote a title for the graph. Next, I put the points on the graph. I found the correct number on the scale for each day, and then followed across to the line for that day. Where the lines met, I put a point. Once I had all the points on the graph, I connected them with lines. I looked for the day that had the point in the highest spot. It was on Friday. So, Friday was the day on which the greatest number of students bought lunch.

Solve a Problem

Studying the problem Read the problem. As you read, think about how you could use a line graph to solve the problem.

On Monday, there was a snowstorm in George's town. Then it warmed up. George decided to measure the snow at 8:00 P.M. each day to see how fast the snow melted. By Monday, there were 10 inches of snow. By Tuesday, the snow measured 7 inches. By Wednesday, there were 6 inches of snow on the ground. George measured 4 inches of snow on Thursday and the same amount on Friday. How much snow melted each day?

Finding the solution Use the information from the problem to complete this line graph. Then write your answer below.

Snow in George's Town

Amount of Snow on Ground (in inches)

Day

Answer: _____

116

Explaining the solution

Reread "Understanding the solution" on page 115, which tells how Nita used a line graph to solve a problem. Then write your own explanation of how you completed the line graph on page 116 and found your solution.

_____

_____

_____

_____

_____

_____

_____

_____

Applying the solution

Use your line graph on page 116 to answer these questions.

1. How deep was the snow on Monday? _____

2. On which two days was the snow the same depth?

   _____

3. How much snow had melted by Wednesday? _____

4. How much snow had melted by Friday? _____

5. By studying the graph, what day would you guess was the warmest? Why do you think so?

   _____

   _____

Learn About Charts

Thinking about the strategy

Students in Mrs. Lucia's class went on a field trip to the Discovery Museum. Mrs. Lucia asked the students to write on a piece of paper what exhibit they liked best. The data is shown below.

Dinosaurs, Electricity, Machines, Dinosaurs, Animals, Dinosaurs, Animals, Machines, Animals, Electricity, Animals, Machines, Dinosaurs, Electricity, Animals, Dinosaurs, Animals, Animals, Machines, Dinosaurs, Animals, Machines, Dinosaurs

A **chart** is a graphic organizer that you can use to show amounts. How can the students find the exhibit that got the most votes? They used this chart to find the answer.

Favorite Exhibits at the Discovery Museum	
Exhibit	**Number of Votes**
Animals	8
Dinosaurs	7
Electricity	3
Machines	5

Answer: _The animal exhibit got the most votes._

Understanding the solution

Read what one of Mrs. Lucia's students wrote to explain how he used a chart to solve the problem.

Find how many columns are needed.

Find how many rows are needed.

Be sure to count each piece of data.

I studied the data and saw that there were 4 different exhibits listed. That means the chart must have 5 rows. The top row is for the titles. The chart must have 2 columns. One column is for the name of the exhibit. The second column is for the number of votes. I wrote the labels at the top of the chart and listed the exhibit names. Next, I counted the votes for each exhibit. I wrote the numbers in the second column, next to the names. I also wrote a title for the chart. By comparing the numbers, I could tell that the animal exhibit got the most votes.

Solve a Problem

Finding the solution

Marta's science class is learning about trees. The class split into four groups. Each group took a section of the town park. They counted the different kinds of trees they saw in their section. Their data is shown below. The park has the most of what kind of tree?

Complete this chart to show the numbers for each kind of tree. Then write your answer below.

Group 1: 4 oak, 12 maple, 3 pine, 8 other
Group 2: 8 maple, 2 oak, 15 pine, 4 other
Group 3: 5 other, 6 pine, 5 maple, 10 oak
Group 4: 2 maple, 5 pine, 10 other, 6 oak

Kind of Tree	Group 1	Group 2	Group 3	Group 4	Total
Maple					
Oak					
Pine					
Other					

Answer: _____

Explaining the solution

Write an explanation of how you found the answer by completing the chart above.

119

Numbers in Context

Read "Open House at Camp Kodiak." Think about the ways that numbers are used in the selection. Then answer items A–C on page 121.

Open House at Camp Kodiak

Camp Kodiak is a summer camp on Big Bear Lake. Many children enjoy spending time there during summer vacation. They get to play sports, do crafts, swim, boat, and make friends.

On the first day of camp, there is an open house. Parents can spend the day at the camp. They take a tour of the grounds. They can even take part in the activities that the children will do during their stay.

Parents can check into the camp with their children between 8:00 A.M. and 12:00 P.M. The camp director keeps a record of when campers check in. Today, between 8:00 and 9:00, 12 campers checked in. Between 9:00 and 10:00, 15 children arrived at camp. Between 10:00 and 11:00, another 13 campers checked in. From 11:00 to 12:00, 10 more children arrived at Camp Kodiak.

Campers and parents can learn about the different camp activities during the open house. There are two sessions for each activity. The counselors talk about and show what campers will be doing. During session A, counselors counted these numbers of visitors in each activity: crafts—18, boating—8, swimming—10, hiking—15, sports—20. In session B, counselors counted these visitors: hiking—11, boating—13, sports—17, crafts—12, swimming—16.

By the end of the open house, campers were tired and excited. It was going to be a great summer!

A. In what time period did the greatest number of children check in at camp? Use the information from page 120 to complete this line graph. Then write your answer below.

Check-in at Camp Kodiak

[Line graph with y-axis labeled "Number of Children" from 0 to 16, and x-axis labeled "Time Period" with four blank entries]

Answer: _____

B. What activity had the most visitors in sessions A and B together? Use the information from page 120 to complete this chart. Then write your answer below.

Number of Visitors at Open House			
Activity	Session A	Session B	Total
Boating			
Crafts			
Hiking			
Sports			
Swimming			

Answer: _____

C. Explain your solution to either item A or item B above.

_____

_____

_____

_____

121

Check Your Understanding

Fill in the letter of the correct answers to questions 1–8.
Write your answers to questions 9 and 10.

This line graph shows the number of books sold each day at the West School Book Fair. Use the graph to answer questions 1–4.

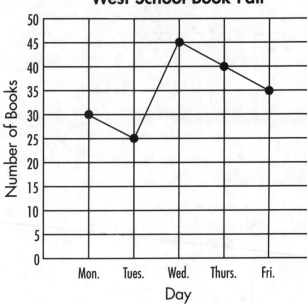

1. On what day were the most books sold?
   - Ⓐ Monday
   - Ⓑ Tuesday
   - Ⓒ Wednesday
   - Ⓓ Thursday

2. How many more books were sold on Monday than on Tuesday?
   - Ⓐ 1 book
   - Ⓑ 5 books
   - Ⓒ 10 books
   - Ⓓ 15 books

3. On what day were the fewest books sold?
   - Ⓐ Friday
   - Ⓑ Thursday
   - Ⓒ Wednesday
   - Ⓓ Tuesday

4. How many books were sold in all?
   - Ⓐ 175 books
   - Ⓑ 170 books
   - Ⓒ 155 books
   - Ⓓ 150 books

Mr. Cruz owns a hardware store. He made this chart to show how many cans of paint he has in the store. Use the chart to answer questions 5–8.

Cans of Paint in Stock		
Kind of Paint	Quarts	Half-gallons
Brand A	12	24
Brand B	10	22
Brand C	26	40
Brand D	16	30

5. How many more half-gallons of brand A are there than quarts of brand A?
   - Ⓐ 24
   - Ⓑ 14
   - Ⓒ 12
   - Ⓓ 10

6. How many quarts of all brands are there?
   - Ⓐ 116 quarts
   - Ⓑ 74 quarts
   - Ⓒ 66 quarts
   - Ⓓ 64 quarts

7. How many more half-gallons of brand C are there than half-gallons of brand D?
   - Ⓐ 1 half-gallon
   - Ⓑ 5 half-gallons
   - Ⓒ 10 half-gallons
   - Ⓓ 70 half-gallons

8. How many cans, both quarts and half-gallons, of brand D are in the store?
   - Ⓐ 46 cans
   - Ⓑ 30 cans
   - Ⓒ 36 cans
   - Ⓓ 14 cans

9. Look at the graph above question 1. At the beginning of the book fair, there were 200 books. At the end of the fair, how many books were left?

10. Kylie wrote down the kinds of vehicles that passed by her house. How many of each kind of vehicle went by? Explain how you found your answer.

    car, truck, van, truck, motorcycle, van, car, car, van, truck, van, car, van, truck, van, car, motorcycle, car, car, truck

Extend Your Learning

- *Class Survey*

    Take a class survey. Have each student write his or her favorite food, sport, or subject, on a piece of paper. Make a chart to show the results.

- *Science: Water Resources*

    Turn on a cold-water tap just enough so that it is dripping. Put a drinking glass under the tap. After 10 minutes have passed, measure (in centimeters) how high the water is in the glass. Measure again after a total of 20 minutes has passed; then after 30 minutes and after 40 minutes. After taking the 4 measurements, make a line graph to show the increases in the height of the water. Ask your teacher to help you find how much water is wasted each year by one dripping faucet.

STRATEGIES ONE–TWELVE REVIEW

Building Number Sense; Using Estimation

Read "Jennifer Learns a Lesson." Think about the ways that numbers are used in the selection. Then answer questions 1 and 2.

Jennifer Learns a Lesson

Jennifer was helping Mr. Ling, her neighbor, rake his lawn. They had been working for about 3 hours. Mr. Ling was paying Jennifer $6.50 for her help.

"You really don't have to pay me, Mr. Ling," Jennifer said.

Mr. Ling smiled. "You should be rewarded for your hard work."

"Well, thanks. I am trying to save up for a special-edition bike helmet. It's so cool, but very expensive. I've got only $7.39 saved. At the rate I'm going, it will take me 10 years to save enough to buy that helmet."

"Where there's a will, there's a way," Mr. Ling said.

"What does that mean?" Jennifer asked.

"It means that if you want something badly enough, you'll figure out a way to get it," Mr. Ling said.

"Even if it seems impossible?"

"Have you ever heard the story of the tortoise and the hare?" Mr. Ling asked.

"Yes," Jennifer said. "Hare and Tortoise have a race. Hare thinks it's impossible for him to lose because he moves so much faster than Tortoise. So Hare stops to take a nap in the middle of the race. Tortoise keeps on going and wins the race."

"So?" Mr. Ling asked.

"So, may I come back next Saturday and help you?" Jennifer asked.

"Of course!"

1. Jennifer and Mr. Ling had been working for about 3 hours. Which of these mixed numbers is closer to 3 than to 2 or to 4?
   - Ⓐ $2\frac{1}{4}$
   - Ⓑ $3\frac{2}{5}$
   - Ⓒ $3\frac{1}{3}$
   - Ⓓ $4\frac{1}{2}$

2. To the nearest dollar, about how much money will Jennifer have saved after Mr. Ling pays her for raking the lawn?

   _____

Applying Addition; Applying Subtraction; Applying Multiplication; Applying Division

Read "The Ancient Pyramids." Think about the ways that numbers are used in the selection. Then answer questions 3–6.

The Ancient Pyramids

The world's oldest stone buildings are the pyramids of ancient Egypt. They were built to protect the bodies of the pharaohs, or kings, of Egypt. The first known pyramid was built about 2650 B.C.

The people who built the pyramids made them by hand. Thousands of years ago, people could not even imagine the machines or tools that builders use today. The Great Pyramid at Giza was 481 feet tall when it was built. Researchers believe that workers took 20 years to build it. Some of its upper stones are now missing, so now the Great Pyramid stands about 30 feet shorter than its original height.

Another amazing stone structure in Egypt is the Great Sphinx. It is 240 feet long and 66 feet high. It has the face of a man and the body of a lion.

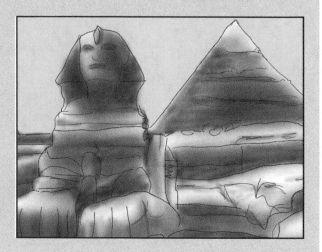

3. If you add 2,650 and 1,999, you will find about how many years ago the first known pyramid was built. How many years ago is that?
   - Ⓐ 4,340 years ago
   - Ⓑ 4,649 years ago
   - Ⓒ 4,659 years ago
   - Ⓓ 3,559 years ago

4. The Statue of Liberty is 305 feet high, from the ground to the top of the statue's head. How much taller was the Great Pyramid when it was built?
   - Ⓐ 176 feet
   - Ⓑ 151 feet
   - Ⓒ 182 feet
   - Ⓓ 200 feet

5. Joey saw an exhibit about ancient Egypt at a museum. In the museum shop, Joey bought 2 picture postcards of the pyramids for each of his 3 sisters. Each postcard cost $1.00. How much did Joey spend?

   _____

6. How many ladders 6 feet high would you have to stack on top of one another to reach the top of the Great Sphinx?

   _____

125

Converting Time and Money; Converting Customary and Metric Measures

Read "Opening Day." Think about the ways that numbers are used in the selection. Then answer questions 7 and 8.

Opening Day

"Hurry up!" Mari called to Shiro. "We're going to be late!"

"You're always in such a rush," Shiro said as he got into the car. He looked at his watch. "It's only 11:53 A.M. The game doesn't start until 1:10 P.M."

"We'll get there on time," Mari's mom said. "I promise."

"Okay," Mari said. "This is not just any game. It's opening day—the first game of the season. I don't want to miss a thing. If we get there early enough, I may be able to get some autographs. Maybe I'll even get to meet Jimmie Lopez."

"Sure," Shiro said. "The best hitter in baseball is going to talk to you."

"It could happen," Mari said.

The ride to the baseball park took 38 minutes. Mari was thrilled when they found their seats. Her seat was just 18 yards from home plate. The pitcher's mound is 60 feet 6 inches from home plate, so Mari was closer to home plate than the pitcher was. She was so close she could hear the player's talking to one another.

Just when Mari was thinking that the day couldn't get any better, it did. Jimmie Lopez hit a foul ball that landed right in Mari's lap. She grabbed the ball and began to jump up and down. Jimmie looked over at Mari and smiled. Then he yelled, "Stay for a few minutes after the game, and I'll sign that for you."

Mari was sure this was the best day of her life!

7. What time did Mari and her family get to the park, and how much time did Mari have to get autographs before the start of the game?

_____

_____

8. Mari's seat was how many inches closer to home plate than was the pitcher's mound?
   - Ⓐ 188 inches
   - Ⓑ 672 inches
   - Ⓒ 78 inches
   - Ⓓ 6 inches

Determining Probability and Averages; Interpreting Graphs and Charts

Read "The Art Experiment." Think about the ways that numbers are used in the selection. Then answer questions 11 and 12. After completing question 12, do "Explaining the Solution."

The Art Experiment

Mrs. D'Alia held up the painting of the bare tree and asked, "Why do you think the artist chose this subject?"

Nicole and the other students couldn't think of a reason. So Mrs. D'Alia placed a daisy, a football, a dented can, an American flag, and a paintbrush on a stand. She asked the students in each of her 4 art classes to choose an object to paint. Later, she created a chart to show the numbers of each object painted.

Object	Class 1	Class 2	Class 3	Class 4
Daisy	5	6	9	3
Football	4	3	1	5
Tin can	0	2	4	3
Flag	7	9	10	8
Paintbrush	2	2	0	1
Number of Students	**18**	**22**	**24**	**20**

Then Mrs. D'Alia asked the students to explain how they had made their choices. Students said they chose an object because of its color, or because it reminded them of a special feeling, or because of its interesting shape. Mrs. D'Alia had helped students understand some reasons that artists choose a subject to paint.

11. What is the average number of students in Mrs. D'Alia's 4 art classes?
    - Ⓐ 20 students
    - Ⓑ 22 students
    - Ⓒ 21 students
    - Ⓓ 23 students

12. What object was chosen least often, and how many times was it chosen?

    _____

    _____

    _____

Explaining the solution Look at your solutions to questions 1–12 on pages 124–128. Choose two questions, and write an explanation of how you found the solution to each. Use a separate sheet of paper.

Using Algebra; Using Geometry

Read "A Rainy Afternoon and a Box of Blocks." Think about the ways that numbers are used in the selection. Then answer questions 9 and 10.

A Rainy Afternoon and a Box of Blocks

Tyler and his friend Otis found a huge box of blocks in the storage room of Tyler's building. "Where did all these blocks come from?" Otis asked.

"My dad made those for me when I was little," Tyler said. "He was taking a woodworking class. He got really good at making wooden blocks."

"I'll say," Otis said. "Let's take the box upstairs. We can build a city of skyscrapers."

Tyler and Otis carried the box upstairs to Tyler's apartment. They separated the blocks by size. There were 5 different-sized blocks. The top of the smallest block was 2 inches long and 3 inches wide. The top of the largest size was 8 inches long and 4 inches wide.

The boys made a grid to show where they would place each skyscraper in their city.

Tyler and Otis spent the whole afternoon creating their city of blocks. Later, Otis said, "I thought blocks were just for babies, but I had fun. Do you think your dad would make me some blocks?"

"We can ask," Tyler said. "But I have to warn you: once he gets started again, he may never stop."

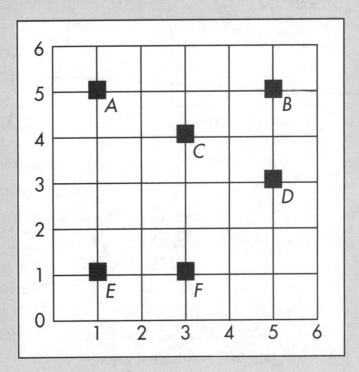

9. What coordinate pair names the location of skyscraper C?
   Ⓐ (3, 4)   Ⓒ (3, 3)
   Ⓑ (4, 5)   Ⓓ (4, 3)

10. What is the sum of the area of the top of the smallest block and the area of the top of the largest block?